Lecture Notes in Mechanical Engineering

Lecture Notes in Mechanical Engineering (LNME) publishes the latest developments in Mechanical Engineering—quickly, informally and with high quality. Original research reported in proceedings and post-proceedings represents the core of LNME. Volumes published in LNME embrace all aspects, subfields and new challenges of mechanical engineering. Topics in the series include:

- Engineering Design
- Machinery and Machine Elements
- Mechanical Structures and Stress Analysis
- Automotive Engineering
- Engine Technology
- Aerospace Technology and Astronautics
- Nanotechnology and Microengineering
- Control, Robotics, Mechatronics
- MEMS
- Theoretical and Applied Mechanics
- Dynamical Systems, Control
- Fluid Mechanics
- Engineering Thermodynamics, Heat and Mass Transfer
- Manufacturing
- Precision Engineering, Instrumentation, Measurement
- Materials Engineering
- Tribology and Surface Technology

To submit a proposal or request further information, please contact the Springer Editor of your location:

China: Ms. Ella Zhang at ella.zhang@springer.com
India: Priya Vyas at priya.vyas@springer.com
Rest of Asia, Australia, New Zealand: Swati Meherishi at swati.meherishi@springer.com
All other countries: Dr. Leontina Di Cecco at Leontina.dicecco@springer.com

To submit a proposal for a monograph, please check our Springer Tracts in Mechanical Engineering at https://link.springer.com/bookseries/11693 or contact Leontina.dicecco@springer.com

Indexed by SCOPUS. All books published in the series are submitted for consideration in Web of Science.

More information about this series at https://link.springer.com/bookseries/11236

Filip Gorski · Michał Rychlik ·
Răzvan Păcurar
Editors

Advances in Manufacturing III

Volume 5 - Biomedical Engineering: Research and Technology Innovations, Industry 4.0

Springer

Editors
Filip Gorski
Faculty of Mechanical Engineering
Poznań University of Technology
Poznan, Poland

Michał Rychlik
Faculty of Mechanical Engineering
Poznań University of Technology
Poznan, Poland

Răzvan Păcurar
Faculty of Machine Building
Technical University of Cluj-Napoca
Cluj-Napoca, Romania

ISSN 2195-4356 ISSN 2195-4364 (electronic)
Lecture Notes in Mechanical Engineering
ISBN 978-3-030-99768-7 ISBN 978-3-030-99769-4 (eBook)
https://doi.org/10.1007/978-3-030-99769-4

This Springer imprint is published by the registered company Springer Nature Switzerland AG
The registered company address is: Gewerbestrasse 11, 6330 Cham, Switzerland

Preface

This volume of Lecture Notes in Mechanical Engineering gathers selected papers presented at the 7th International Scientific-Technical Conference MANUFACTURING 2022, held in Poznan, Poland, on May 16–19, 2022. The conference was organized by the Faculty of Mechanical Engineering, Poznan University of Technology, Poland.

The aim of the conference was to present the latest achievements in the broad field of mechanical engineering and to provide an occasion for discussion and exchange of views and opinions. The conference covered topics as follows:

- mechanical engineering
- production engineering
- quality engineering
- measurement and control systems
- biomedical engineering.

The organizers received 165 contributions from 23 countries around the world. After a thorough peer-review process, the committee accepted 91 papers for conference proceedings prepared by 264 authors from 23 countries (acceptance rate around 55%). Extended versions of selected best papers will be published in the following journals: *Management and Production Engineering Review, Bulletin of the Polish Academy of Sciences: Technical Sciences, Materials, Applied Sciences.*

The book **Advances in Manufacturing III** is organized into five volumes that correspond to the main conference disciplines mentioned above.

Advances in Manufacturing III – Volume 5 - Biomedical Engineering: Research and Technology Innovations, Industry 4.0 presents current research and technology trends in the young, yet exciting field of biomedical engineering. Chapters were carefully selected to highlight important advancements in the field of design, digital analysis, and 3D printing of medical devices. Current challenges related to the application of medical imaging and 3D scanning are discussed, as well as the application of finite element analysis in bioimaging. Additive manufacturing technologies used for developing personalized medical devices are also discussed. Further, the application of production and quality management tools in medical engineering is described. This book gathers 13 chapters prepared by 40 authors from five countries.

We would like to thank the members of the international program committee for their hard work during the review process.

We acknowledge all people that contributed to the staging of MANUFACTURING 2022: authors, committees, and sponsors. Their involvement and hard work were crucial to the success of the MANUFACTURING 2022 conference.

May 2022

Filip Górski
Michał Rychlik
Răzvan Păcurar

Organization

Steering Committee

General Chair

Adam Hamrol	Poznan University of Technology, Poland

Chairs

Olaf Ciszak	Poznan University of Technology, Poland
Magdalena Diering	Poznan University of Technology, Poland
Justyna Trojanowska	Poznan University of Technology, Poland

Scientific Committee

Ahmad Majdi Abdul-Rani	University of Technology PETRONAS, Malaysia
Dario Antonelli	Politecnico di Torino, Italy
Katarzyna Antosz	Rzeszow University of Technology, Poland
Jorge Bacca-Acosta	Fundación Universitaria Konrad Lorenz, Colombia
Zbigniew Banaszak	Koszalin University of Technology, Polska
Andre D. L. Batako	Liverpool John Moores University, UK
Stefan Berczyński	West Pomeranian University of Technology in Szczecin, Poland
Alain Bernard	Ecole Centrale de Nantes, France
Kristina Berladir	Sumy State University, Ukraine
Christopher Brown	Worcester Polytechnic Institute, USA
Anna Burduk	Wrocław University of Science and Technology, Poland
Marcin Butlewski	Poznan University of Technology, Poland
Lenka Čepová	VSB Technical University of Ostrava, Czech Republic
Nadežda Čuboňová	University of Žilina, Slovakia

Somnath Chattopadhyaya	Indian School of Mines, India
Olaf Ciszak	Poznań University of Technology, Polska
Jacek Diakun	Poznań University of Technology, Poland
Magdalena Diering	Poznan University of Technology, Poland
Ewa Dostatni	Poznan University of Technology, Poland
Jan Duda	Cracow University of Technology, Poland
Davor Dujak	J.J. Strossmayer University of Osijek, Croatia
Milan Edl	University of West Bohemia, Czech Republic
Brigita Gajšek	University of Maribor, Slovenia
Mosè Gallo	University of Naples Federico II, Italy
Adam Gałuszka	Silesian University of Technology, Poland
Józef Gawlik	Cracow University of Technology, Poland
Adam Gąska	Cracow University of Technology, Poland
Adam Górny	Poznan University of Technology, Polska
Filip Górski	Poznan University of Technology, Poland
Adam Hamrol	Poznan University of Technology, Poland
Mukund Harugade	Padmabhooshan Vasantraodada Patil Institute of Technology, India
Ivan Hudec	Slovak University of Technology in Bratislava, Slovakia
Jozef Husár	Technical University of Košice, Slovakia
Aminul Islam	Technical University of Denmark, Denmark
Vitalii Ivanov	Sumy State University, Ukraine
Andrzej Jardzioch	West Pomeranian University of Technology in Szczecin, Poland
Sabahudin Jašarević	University of Zenica, Bosnia and Herzegovina
Małgorzata Jasiulewicz-Kaczmarek	Politechnika Poznańska, Poland
Wojciech Kacalak	Technical University of Koszalin, Poland
Sławomir Kłos	University of Zielona Góra, Polska
Lucia Knapčíková	Technical University of Košice, Slovakia
Damian Krenczyk	Silesian University of Technology, Poland
Jolanta Krystek	Silesian University of Technology, Poland
Józef Kuczmaszewski	Lublin University of Technology, Poland
Agnieszka Kujawińska	Poznan University of Technology, Poland
Ivan Kuric	University of Žilina, Slovak Republic, Slovak Republic
Piotr Łebkowski	AGH, University of Science and Technology, Poland
Stanisław Legutko	Poznan University of Technology, Polska
Sławomir Luściński	Kielce University of Technology, Poland
José Machado	University of Minho, Portugal
Vijaya Kumar Manupati	National Institute of Technology, Warangal, India
Uthayakumar Marimuthu	Kalasalingam Academy of Research and Education, India

Jorge Martín Gutiérrez	Universidad de La Laguna, Spain
Józef Matuszek	University of Bielsko-Biala, Poland
Arkadiusz Mężyk	Silesian University of Technology, Poland
Dariusz Mikołajewski	Kazimierz Wielki University, Poland
Andrzej Milecki	Poznań University of Technology, Poland
Piotr Moncarz	Stanford University, USA
Manurl Francisco Morales-Contreras	Universidad Pontificia Comillas, Spain
Keddam Mourad	USTHB, Algeria
Leticia Neira	UANL, México
Przemysław Niewiadomski	University of Zielona Góra, Polska
Erika Ottaviano	University of Cassino and Southern Lazio, Cassino
Eren Özceylan	Gaziantep University, Turkey
Răzvan Păcurar	Technical University of Cluj-Napoca, Romania
Justyna Patalas-Maliszewska	University of Zielona Góra, Poland
Ivan Pavlenko	Sumy State University, Ukraine
Lucjan Pawłowski	Lublin University of Technology; Polish Academy of Science, Poland
Dragan Peraković	University of Zagreb, Croatia
Alejandro Pereira	Universidade de Vigo, Spain
Marko Periša	University of Zagreb, Croatia
Pierluigi Rea	University of Cagliari, Italy
Ján Piteľ	Technical University of Košice, Slovakia
Izabela Rojek	Kazimierz Wielki University, Poland
Alessandro Ruggiero	University of Salerno, Italy
Dominik Rybarczyk	Poznań University of Technology, Polska
Krzysztof Santarek	Warsaw University of Technology, Poland
Jarosław Sęp	Rzeszow University of technology, Poland
Bożena Skołud	Silesian University of Technology, Polska
Jerzy Andrzej Sładek	Cracow University of Technology, Poland
Dorota Stadnicka	Rzeszów University of Technology, Poland
Beata Starzyńska	Poznan University of Technology, Poland
Manuel Francisco Suárez Barraza	Universidad de las Américas Puebla, Mexico
Marcin Suszyński	Poznan University of Technology, Polska
Marek Szostak	Poznan University of Technology, Poland
Erfan Babaee Tirkolaee	Istinye University, Turkey
Justyna Trojanowska	Poznan University of Technology, Poland
Leonilde Varela	University of Minho, Portugal
Katarzyna Węgrzyn-Wolska	Engineering School of Digital Technologies, France
Gerhard-Wilhelm Weber	Poznan University of Technology, Poland
Edmund Weiss	Calisia University, Poland
Dorota Więcek	University of Bielsko-Biala, Poland

Michał Wieczorowski	Poznan University of Technology, Poland
Hanna Włodarkiewicz-Klimek	Poznan University of Technology, Poland
Szymon Wojciechowski	Poznan University of Technology, Poland
Ralf Woll	BTU Cottbus-Senftenberg, Germany
Jozef Zajac	Košice University of Technology, Slovakia
Nermina Zaimovic-Uzunovic	University of Zenica, Bosnia and Herzegovina

Program Committee

Karla Alvarado-Ramírez
José Angel
Dario Antonelli
Katarzyna Antosz
Erfan Babaee Tirkolaee
Jorge Bacca-Acosta
Diana Baila
Prashanth Bandari
Lucía Barcos-Redín
Andre Batako
Petr Beneš
Kristina Berladir
Marcin Białek
Mikołaj Bilski
Matej Borovinšek
Anna Borucka
Cristina Borzan
Sara Bragança
Daniel Brissaud
Christopher Brown
Paweł Buń
Anna Burduk
Bartłomiej Burlaga
Jacek Buśkiewicz
Marcin Butlewski
Massimiliana Carello
Fernando Castillo
Robert Cep
Miroslav Cisar
Olaf Ciszak
Felipe Contreras
Tomáš Coranič
Eric Costa
Margareta Coteata
Dorota Czarnecka-Komorowska
Reggie Davidrajuh
Mario del Valle
Boris Delibašić
Jakub Demčák
Jacek Diakun
Magdalena Diering
Grzegorz Domek
Ewa Dostatni
Radosław Drelich
Jan Duda
Luboslav Dulina
Eduard Franas
Paweł Fritzkowski
Brigita Gajšek
Mosè Gallo
Bartosz Gapiński
Katarzyna Gawdzińska
Adam Gąska
Dominik Gojdan
Arkadiusz Gola
Fernando González-Aleu
Adam Górny
Filip Górski
Cezary Grabowik
Marta Grabowska
Jakub Grabski
Damian Grajewski
Patrik Grznár
Aleksander Gwiazda
Michal Hajžman
Adam Hamrol
Mukund Harugade
Amalija Horvatić-Novak

Stella Hrehová
Jozef Husár
Vitalii Ivanow
Carmen Jaca
Rajat Jain
Michał Jakubowicz
Małgorzata Jankowska
Andrzej Jardzioch
Małgorzata Jasiulewicz-Kaczmarek
Anna Karwasz
Jakub Kaščak
Sławomir Kłos
Lucia Knapčíková
Adam Koliński
Elif Kongar
Boris Kostov
Kateryna Kovbasiuk
Arkadiusz Kowalski
Tomasz Kowaluk
Damian Krenczyk
Grzegorz Królczyk
Józef Kuczmaszewski
Agnieszka Kujawińska
Panagiotis Kyratsis
Georgios Lampropoulos
Peter Lazorík
Stanislaw Legutko
Andrzej Loska
Czesław Łukianowicz
José Machado
Marek Macko
Ján Majerník
Vitalii Maksymenko
Damjan Maletič
Matjaž Maletič
Vijaya Kumar Manupati
Sven Maricic
Jorge Martín Gutiérrez
Thomas Mathia
Maciej Matuszewski
Dariusz Mazurkiewicz
Jakub Michalski
Dariusz Mikołajewski
Janusz Mleczko
Ladislav Morovic
Keddam Mourad
Leticia Neira
Zdeněk Neusser
Magdalena Niemczewska-Wójcik
Przemysław Niewiadomski
Czesław Niżankowski
Nejc Novak
Filip Osiński
Erika Ottaviano
Eren Ozceylan
Ancuta Pacurar
Kyratsis Panagiotis
Jason Papathanasiou
Waldemar Paszkowski
Justyna Patalas-Maliszewska
Ivan Pavlenko
Paweł Pawlus
Jarosław Pempera
Dragan Peraković
Alejandro Pereira
Dariusz Plinta
Florin Popister
Evangelos Psomas
Paulina Rewers
Francisco Gabriel Rodriguez-González
Michał Rogalewicz
Izabela Rojek
Mirosław Rucki
Biserka Runje
Dominik Rybarczyk
Michał Rychlik
Milan Saga
Alžbeta Sapietova
Filip Sarbinowski
Holger Schlegel
Armin Schleinitz
Dariusz Sędziak
Robert Sika
Zbyněk Šika
Varinder Singh
Bożena Skołud
António Lucas Soares
Ewa Stachowska
Dorota Stadnicka
Sergiu Dan Stan

Roman Starosta
Beata Starzyńska
Krzysztof Stępień
Tomasz Stręk
Manuel Suárez-Barraza
Marcin Suszynski
Grażyna Sypniewska-Kamińska
Marek Szostak
Jiří Tengler
Samala Thirupathi
Francisco Torres
Justyna Trojanowska
Piotr Trojanowski
Krzysztof Tyszczuk
M. Uthayakumar
Leonilde Varela
Nikola Vitkovic
Adrián Vodilka
Sachin Waigaonkar
Tomasz Walczak
Łukasz Warguła
Gerhard-Wilhelm Weber
Katarzyna Węgrzyn-Wolska
Radosław Wichniarek
Michał Wieczorowski
Dariusz Więcek
Hanna Włodarkiewicz-Klimek
Szymon Wojciechowski
Adam Woźniak
Ryszard Wyczółkowski
Nermina Zaimovic-Uzunovic
Ivan Zajačko
Magdalena Żukowska
Krzysztof Żywicki

Special Sessions

Manufacturing and Management Approaches and Tools for Companies in Era of Digitalization

Special Session Organizing Committee

Leonilde Varela	University of Minho, Portugal
Justyna Trojanowska	Poznan University of Technology, Poland
Vijaya Kumar Manupati	Mechanical Engineering Department, India
Paulina Rewers	Poznan University of Technology, Poland

Operational Innovation and Excellence in Organizations of Twenty-First Century. Lean Thinking, Kaizen Philosophy, and Toyota Production System

Manuel F. Suárez-Barraza	Universidad de las Américas Puebla, Mexico
Manuel F. Morales-Contreras	Universidad Pontificia Comillas, Spain
Agnieszka Kujawińska	Poznan University of Technology, Poland
Marta Grabowska	Poznan University of Technology, Poland

Cyber-Physical Production Systems: From Design to Applications

Katarzyna Antosz	Rzeszow University of Technology, Poland
José Machado	University of Minho, Portugal
Erika Ottaviano	University of Cassino and Southern Lazio, Italy
Pierluigi Rea	University of Cagliari, Italy

Augmented, Virtual and Mixed Reality Systems

Jozef Husár	Technical University of Košice, Slovak Republic
Jakub Kaščak	Technical University of Košice, Slovak Republic
Paweł Buń	Poznan University of Technology, Poland

Intelligent Methods Supporting Manufacturing Systems Efficiency

Anna Burduk	Wroclaw University of Science and Technology, Poland
Andre Batako	Liverpool John Moores University, UK
Dorota Więcek	University of Bielsko-Biala, Poland
Ivan Kuric	University of Žilina, Slovak Republic

Sustainability on Production in the Aspect of Industry 4.0

Ewa Dostatni	Poznan University of Technology, Poland
Izabela Rojek	Kazimierz Wielki University in Bydgoszcz, Poland
Jacek Diakun	Poznan University of Technology, Poland
Dariusz Mikołajewski	Kazimierz Wielki University in Bydgoszcz, Poland

Design and Rapid Manufacturing of Customized Medical Products

Filip Górski	Poznan University of Technology, Poland
Magdalena Żukowska	Poznan University of Technology, Poland
Răzvan Păcurar	Technical University of Cluj-Napoca, Romania

Metrology and Statistical Analysis of Measurement and Control Systems

Alejandro Pereira Domínguez	University of Vigo, Spain
Lenka Čepová	VSB - Technical University of Ostrava, Czech Republic
Magdalena Diering	Poznan University of Technology, Poland
Bartosz Gapiński	Poznan University of Technology, Poland

Analytical and Computational Methods in (Bio-) Mechanical and Material Engineering

Tomasz Stręk	Poznan University of Technology, Poland
Roman Starosta	Poznan University of Technology, Poland
Nejc Novak	University of Maribor, Slovenia
Pavel Polach	University of West Bohemia, Czech Republic

Contents

Evaluation of a Prototype System of Automated Design and Rapid Manufacturing of Orthopaedic Supplies

Filip Górski(✉), Radosław Wichniarek, Wiesław Kuczko, Magdalena Żukowska, Justyna Rybarczyk, and Monika Lulkiewicz

Faculty of Mechanical Engineering, Poznan University of Technology, Piotrowo 3, 60-965 Poznań, Poland
filip.gorski@put.poznan.pl

Abstract. The aim of this study was to evaluate the AutoMedPrint system, which is used for automated design and rapid production of orthopaedic and prosthetic devices based on anthropometric measurements. The system was tested on a group of 21 patients. Research has been conducted to determine their needs and expectations. In the first step, anthropometric data was collected using a 3D scanner. The obtained data allowed to automatically design wrist-hand orthosis using an intelligent CAD model. Patients customized their orthoses using interactive product configurator available in the system. Some orthoses were manufactured using the necessary machines by 3D printing, and then handed over to selected patients for their opinion. The paper presents the results of usability tests and the evaluation of the technical aspects of the system. Based on these results, the main problems were characterized and ways to improve the system were suggested. Modifications are designed to improve patient satisfaction, usability, ease of use, and the enjoyment of interacting with a system or product.

Keywords: Additive manufacturing · Customization · Design automation · Medical 3D printing

1 Introduction

The rapid development of additive manufacturing technology in recent years has allowed a significant reduction in costs and time needed to implement a new product. Additive manufacturing processes make it possible to obtain physical, 3D shapes of almost any complexity, directly from a digital representation of the product (usually a model made in a Computer Aided Design - CAD system) [1]. There is no need to use any specialized tooling besides the production machine equipment. These technologies are invaluable when there is a need to quickly manufacturing of a physical prototype of a designed part [1], which is especially important in personalized medicine [2–4]. Using of 3D technology also allows to show the expected product to the patient before its production and provide him/her with flexible design options, which can be used in the process of designing and rapid production of orthopaedic devices. This allows the patient to be

F. Gorski et al. (Eds.): MANUFACTURING 2022, LNME, pp. 1–15, 2022.
https://doi.org/10.1007/978-3-030-99769-4_1

involved in the process, improves communication with the patient, and also allows to receive feedback about the device during the design phase, even before manufacturing. This makes it easier to improve the product in order to increase the satisfaction of its user. The 3D printing processes can also be useful in the field of foods and nutrition [5], patient education [6] and teaching of resident physicians [7].

A wide spectrum of additive manufacturing varieties makes it possible to manufacture products from many types of materials [8, 9]. Additive manufacturing in relation to traditional technologies (casting, machining and plastic molding) has certain limitations related to the efficiency, quality and physicochemical properties of the manufactured products [10]. However, as of 2019, the production of finished parts is much higher than in previous years [11]. One of the most widely used additive manufacturing technologies for industrial purposes is fused deposition modeling (FDM), which can be used to obtain parts from thermoplastic materials. The most commonly used building materials are acrylonitrile butadiene styrene (ABS) and polylactic acid (PLA), which provide relatively good strength and acceptable thermal shrinkage and allow for further processing of the obtained elements. From year to year, the range of available materials that can be processed with the FDM method is constantly growing [9]. FDM machines are quiet and clean, and compared to other additive manufacturing technologies, they have small dimensions and are easy to maintain, so they can be used in design offices, hospitals or medical facilities [2, 3, 12].

One of the most popular classes of 3D printed medical devices in wide use are orthopaedic supplies, primarily limb orthoses, which are medical devices designed to maintain the rigidity and safety of a selected part of the patient's body during healing or convalescence [13–15]. This is mostly achieved by immobilizing and protecting the area around the joint from deformation and physical damage. Orthoses can also be used to force a specific position and mutual orientation of different body parts [16]. The limb orthoses might be universal, which are relatively inexpensive, or personalized, much better but also a much more expensive product and made based on the patient's anatomical measurement [17].

One of the biggest problems with 3D printing custom orthopaedic supplies is engineering expertise. In the modern design process, the patient's anthropometric data are collected and digitally processed, mostly manually, which can generate many inaccuracies [20]. Many hours of advanced surface modeling in CAD systems are required to obtain the shape [18]. 3D-printing of thermoplastic products with satisfactory accuracy and strength values is difficult. The process parameters significantly affect the properties of the obtained parts [21]. Consequently, traditional plaster casting processes have not been replaced by 3D printing yet. Research is regularly conducted to facilitate data collection, processing and production in general medical practice [18, 19, 22].

Due to the problems in the subject of traditional manufacturing of individualised orthopaedic supplies, a completely new automated system, under the name of AutoMed-Print (shortly AMP) has been developed, allowing for automatic design and manufacture of orthopaedic products - limb orthoses and upper limb prostheses. This study was conducted to evaluate an upper limb orthosis using this system with a group of twenty-one people, to present detailed problems regarding the use of the system from the user-patient and user-operator perspectives, and to propose solutions to these problems and other improvements to the system [23].

2 Materials and Methods

2.1 The AutoMedPrint System

The AutoMedPrint system is used for automatic design and additive manufacturing of selected orthoses and prostheses based on patient anthropometric measurements. The system (Fig. 1) consists of a station for 3D scanning and design, a user interface station with applications supporting the scanning process and product configuration, and a station for rapid manufacturing.

Fig. 1. The AutoMedPrint system prototype, a) 3D scanning rig; b) operator interface; c) user interface; station for 3D printing not shown

The design and manufacturing of products is based on data from the process of scanning the patient's limb. Not counting the time it takes to make the product, the whole process takes up to several dozen minutes. The time required to produce a finished product can take up to several dozen hours, depending on the type of product. The system allows defining the type of product the patient needs, taking anthropometric measurements using a non-contact 3D scanning technique, automatically designing the product based on the patient's anthropometric data, designing and visualizing the product by the recipient, and preparing and executing the rapid manufacturing process. The system's scheme of operation is shown in Fig. 2.

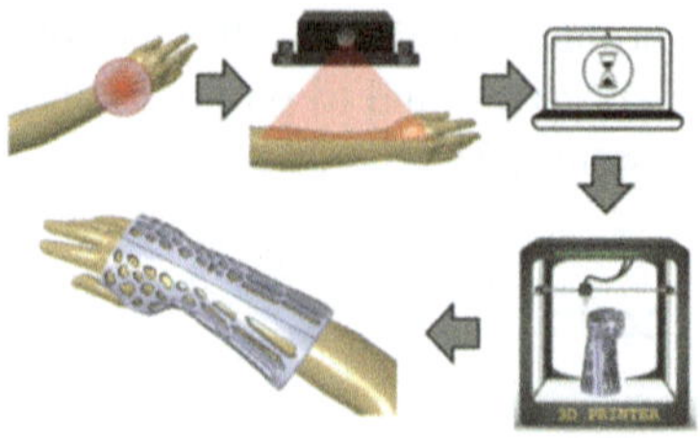

Fig. 2. Scheme of work of the proposed AutoMedPrint system [24]

2.2 Plan of the Experiments

The evaluation of the prototype AMP system was a part of its studies, performed within a scope of an R&D project. Full detailed description is presented in the work [23].

In the first stage of the research, a plan was developed, in which a method for conducting usability tests with users of the AutoMedPrint system was selected and a test course was designed, consisting of the stages of 3D scanning, design and rapid manufacturing. In the next step, a set of questionnaires evaluating particular parts of the system was developed and a leaflet was designed, handed out to the participants beforehand. Then a test group of 21 people was assembled to participate in testing the system and tests were conducted on them in the laboratory, according to the designed plan.

In the next stage, for 6 selected patients, orthoses were manufactured using 3D printing. Once all stages of the study have been completed, survey results were collected and compiled. This allowed to evaluate the system in division into the 3D scanning process, the design process, the rapid manufacturing process, the user interface and the finished product. In the final step, an overall assessment of the whole system was performed.

2.3 Methodology of Experiments

The first step of the study on the evaluation of the manufacturing system was to invite the patient to the laboratory and then hand out an information leaflet on each of the stages of testing the AutoMedPrint system. Once the participant was familiar with the study, the facilitator introduced the different parts of the system and discussed the tasks awaiting the patient and answered any questions.

The next step was the 3D scanning process. Before starting the process, it was necessary to complete the basic data on the patient. The scanning process consisted of two parts. In the first part, the participant uses an interactive application of scanning assistant, providing visual, animated information about the 3D scanning process. In the second part, the participant's upper limb was measured using a 3D scanner to create a 3D model and then a CAD model of the wrist-hand orthosis. If there was a mistake in the 3D scanning process that prevented correct results from being obtained, the process was repeated.

In the next step the participant, in accordance with his preferences, personalized his product through another interactive application – orthosis configurator. The data of the designed product was saved in the system, in a text file. Meanwhile, the operator initiated and overviewed the automated processing of 3D scans, to first obtain a 3D model of the upper limb and then of the orthosis.

Once the 3D scanning and orthotic design processes were completed, the lab testing came to an end and the participant was asked to complete a survey regarding their evaluation of the AutoMedPrint system. The survey was divided into four parts and dealed with the evaluation of 3D scanning process, evaluation of design process, evaluation of user interface and overall system evaluation – this was filled by all participants. The six participants who received the finished, 3D printed product, completed an additional online survey regarding their evaluation of the manufactured orthosis.

2.4 Methodology of System Evaluation

Methodology of 3D Scanning Process Evaluation

The 3D scanning process was evaluated from a technical point of view and from the perspective of the patient, i.e. the system user. It was checked if the scanning process was successful the first time, i.e. if a set of 3D scans needed for upper limb reconstruction is obtained for all participants, or if it should be repeated e.g. due to incorrect limb positioning during the examination. It was also verified that a correct 3D model of the limb could be created for all participants without errors occurring during the processing of the 3D scans. For the group of six for whom the orthoses were manufactured, 3D models of the limbs were inspected for resulting artifacts to assess the need for additional cleaning of the scans. In order to properly carry out further steps leading to the design of the orthosis, the necessity of removing the resulting artifacts was checked and if there is one, the artifacts are removed.

To evaluate the 3D scanning process, which consisted of using an interactive scanning assistant and measuring the upper limb with a 3D scanner, participants completed a questionnaire in which they answered questions about their familiarity with the 3D body scanning process, their comfort and convenience during the 3D scanning process, and the duration of the process. Below are sample questions from the 3D scanning process evaluation survey:

- Have you ever had a 3D body scan done?
- Did you feel anxious about the process before 3D scanning?
- Was the 3D scanning station (chair, table, armrest) comfortable for you?
- How would you rate the duration of the 3D scanning process?
- Do you find measuring your body with a 3D scanner more convenient than measuring your body with traditional methods?

In addition, the study moderator measured the time each system user used the interactive scanning assistant and each person's 3D scanning time using a stopwatch. These times were then summed to obtain the duration of the entire step of the 3D scanning process.

Methodology of the Design Process Evaluation

The design process was evaluated from a technical perspective as well as from the perspective of the patient, the user of the system. For the six patients for whom the orthoses were fabricated, it was verified that it was possible to correctly extract points from the limb model, generate autogenerated model feed sheets and CAD models of the orthoses in an automated manner.

Evaluation of the orthosis design process was done using a completed questionnaire by the participants who answered four one-choice questions. The following are sample questions from the process evaluation survey:

- Can the configurator be used to design a finished product easily and quickly?
- Is the setup process enjoyable and fun?

By answering the questions, participants indicated on a five-point scale how much they agreed with the statements regarding the use of the configurator, its operation and the ease of configuring the product. During the design of the orthosis, the moderator measured the time the system user used the configurator with a stopwatch and noted in the report whether the study participant completed the device configuration process successfully without his or her major intervention.

Methodology of User Interface Assessment

Patients evaluated the user interface by completing a survey in which they answered four questions. The questions were designed to assess the design of the configurator, the readability of icons and layout, and the ease of navigation when configuring the orthosis. They were constructed to determine the intuitiveness of using the interface. Participants determined using a five-point scale the extent to which they agreed with the interface statements that the layout is clear and understandable, icons are easy to identify and understand, the arrangement of elements on the screen promotes easy navigation, and about the visual appeal of the configurator.

Methodology of Evaluation of Manufacturing Process and Finished Product

The evaluation of the rapid fabrication process checked that orthoses were fabricated, one for each participant in the six-person group for which this stage of the study was scheduled. If an error occurred during the process, this information had to be recorded and the process repeated to produce a functional orthosis. The fabrication times of each of the two parts of the orthosis (upper and lower) were measured. User selections from the configurator that provided guidelines for fabrication included the shape of the openwork, material, colour, print orientation (horizontal/vertical), fabrication strategy, machining (with/without sanding) and assembly (with/without tape).

Patients completed a questionnaire about the finished product. The questions concerned, among others, the compliance of the manufactured product with the design and visualization with the configurator and their expectations, as well as the quality of the manufactured product.

Methodology of Overall System Assessment
In order to evaluate the system in general, participants completed a questionnaire. Sample questions from the survey are shown below:

- How satisfied are you with the services offered by the system?
- Do you find the system helpful when purchasing a personalized orthopaedic device?
- What do you think are the main advantages of the system?
- How would you rate the duration of the entire process - from 3D scanning to product configuration?
- Would you use such a system if the need arose in the future?

Questions were designed to elicit information about patients' overall impressions.

3 Results

3.1 Process Results

3D Scanning Process
Based on the evaluation of the 3D scan, it can be concluded that the process was successful the first time for every patient (21) of the study. During the experiment, 20 out of 21 correct upper limb models were obtained without interfering with the obtained partial scans.

Based on the six limb models selected for the design of the orthosis, it was found that in two patients the positioned hand for scanning was bent - the metacarpus pushed outwards. Care must therefore be taken to ensure that patients position the limb in addition to the correct position of the entire upper limb during the 3D scanning process.

The obtained limb models in most test patients had visible artifacts (Fig. 3), not corresponding to the anatomical structure.

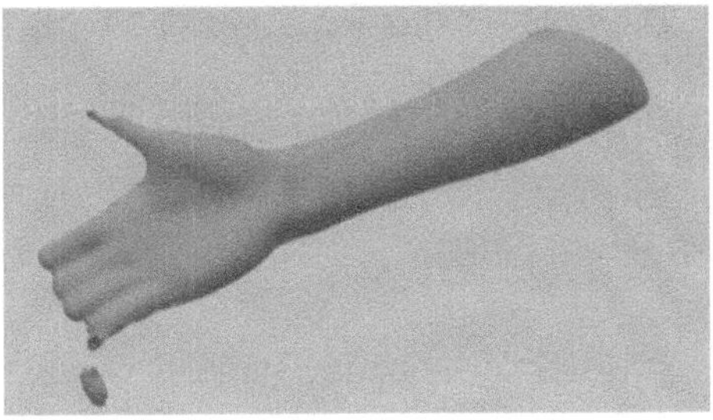

Fig. 3. Scanned right upper limb with artifacts

In order to properly model the orthosis in a further step, artifacts were removed manually using dedicated software. The total average time to use the interactive scanning assistant was 1 min and 12 s.

Design Process

For all selected products, the corresponding points from the cross-sections were generated correctly. For two orthoses, complete sheets feeding the autogenerated model could not be obtained, making it impossible to obtain a correct model in Autodesk Inventor. For the other three orthoses, the model updated in the software with errors. Problems encountered required manual intervention with sheets or models in Inventor. The selected CAD model of the orthosis is shown in Fig. 4.

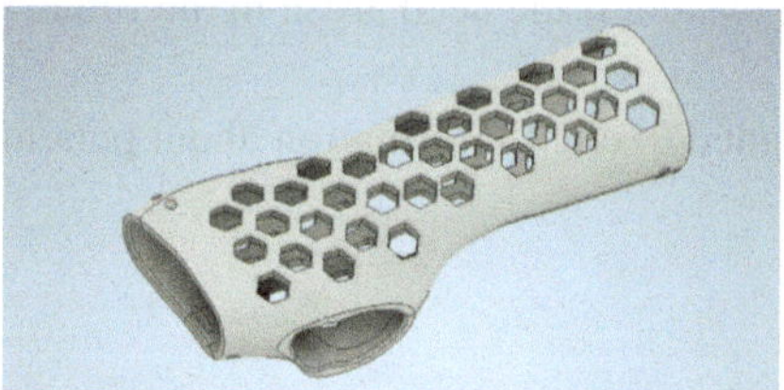

Fig. 4. CAD model of the orthosis [23]

The average design time was 2 min and 2 s. This was the time when patients configured the product according to their preferences for appearance and functionality and the operator received instructions on how to manufacture the orthosis.

On average, the patient used the scanning and design module for 8 min 26 s, and their entire study took an average of 12 min 40 s.

Rapid Manufacturing Process

This step was successfully accomplished for all selected patients, thus six orthoses, 12 parts in total, were made correctly and without errors. The processes were stable and required little operator supervision. After fabrication and basic finishing, it was checked whether it was possible to assemble the two parts together. For each orthosis, the halves fit together, so it can be concluded that the manufactured products fulfilled the role of stiffening the wrist. The orthoses that were provided to the study participants for evaluation are shown in Fig. 5.

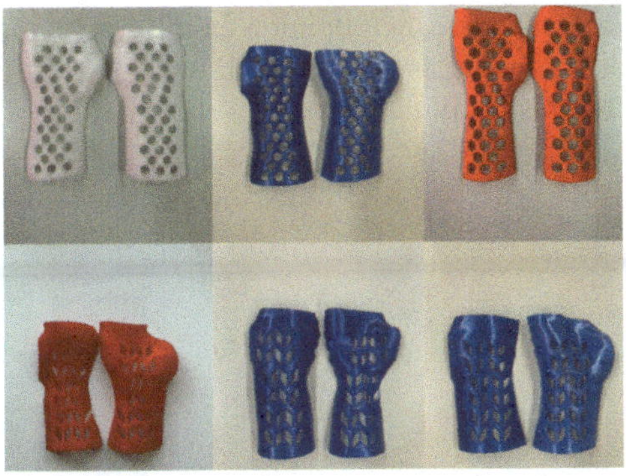

Fig. 5. Manufactured orthoses [23]

3.2 Evaluation Results

3D Scanning Process

According to the patients' responses, most of them (19 patients) had not experienced 3D body scanning before. Although the majority of patients had not encountered 3D scanning in the context of body measurement, none of the study participants had concerns or felt anxious about the process.

The respondents were also asked to rate the duration of the 3D scanning process (everyone was asked to answer according to their feelings). Just over half of the patients (11) rated the duration of the 3D scanning process as short very short (1). For eight people, the scan time was just right, meaning neither too long nor too short.

When asked about the convenience of measuring the body with a 3D scanner than with traditional methods, respondents answered almost unanimously by indicating yes (9) or definitely yes (11) that non-contact scanner measurement is more convenient for them.

The above mentioned results are shown in Fig. 6.

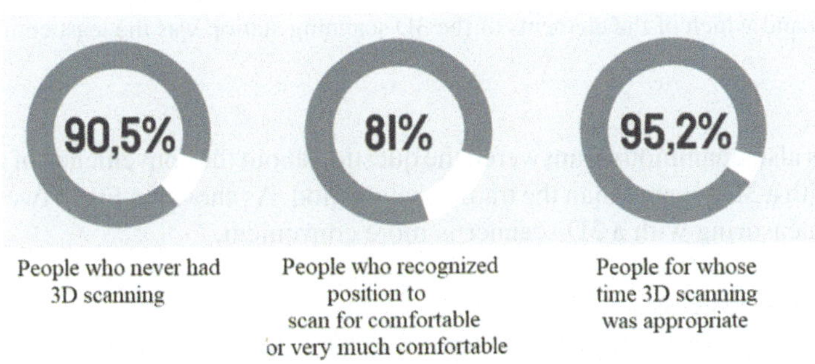

Fig. 6. The results of the 3D scanning process

Patients were also asked to give their opinion on the ergonomics of the workstation, which consists of a chair, a table and an armrest, and to select the least comfortable workstation element. The results of the survey are shown in Fig. 7.

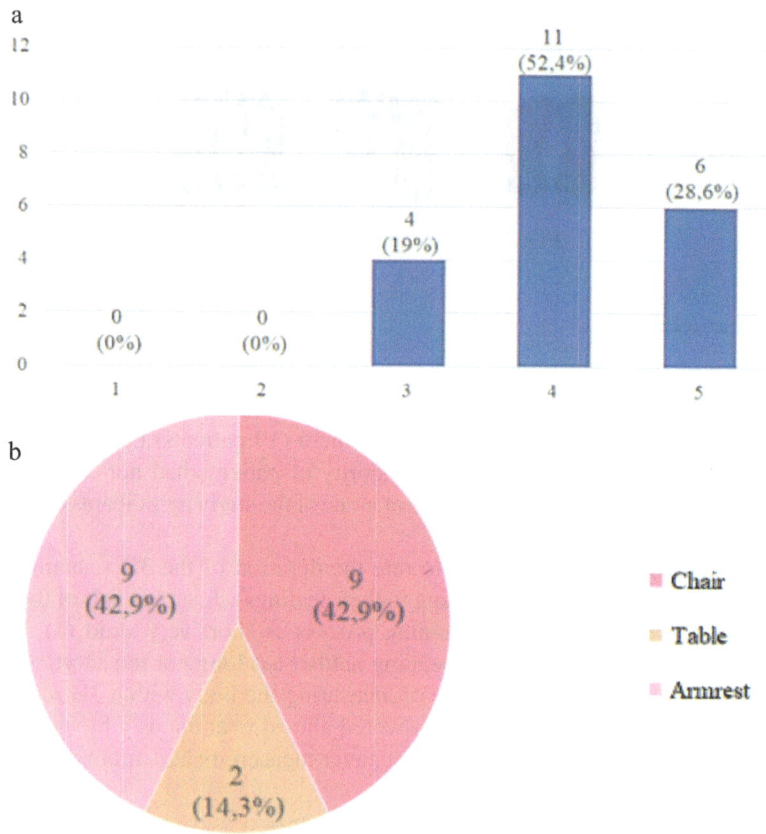

Fig. 7. Answers for questions: Was the 3D scanning station (chair, table, armrest) comfortable for you? (A) and which of the elements of the 3D scanning station was the least comfortable for you? (B)

Patients also unanimously answered the question about the convenience of measuring the body with a 3D scanner than the traditional method. As shown in Fig. 8 twenty people think that measuring with a 3D scanner is more convenient.

Customization Process

The vast majority of patients (20) stated that with the help of the configurator it is possible to easily and quickly design ready orthoses. Seventeen participants agreed that the setup process was fun and enjoyable. The results are shown in Fig. 9.

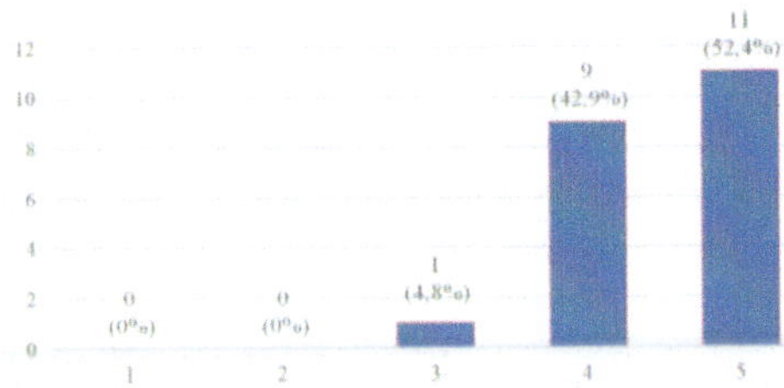

Fig. 8. Answer for question: Is body measurement using the 3D scanner more convenient for you than measuring the body using traditional methods?

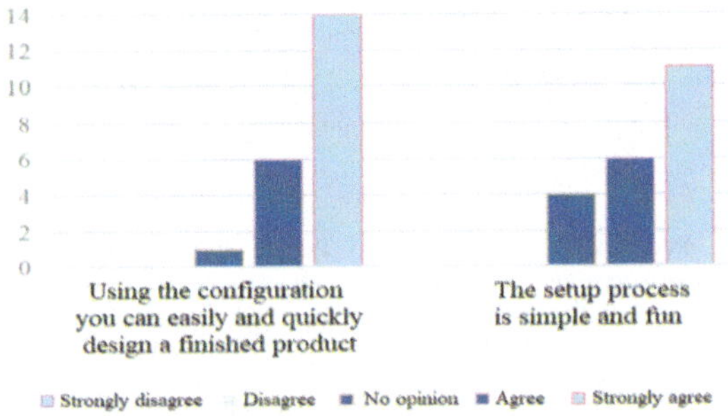

Fig. 9. Results of a survey on the orthosis design process

User Interface Assessment

Based on the responses, it can be seen that a large majority of the respondents (20) believe that the interface is clear and understandable (Fig. 10).

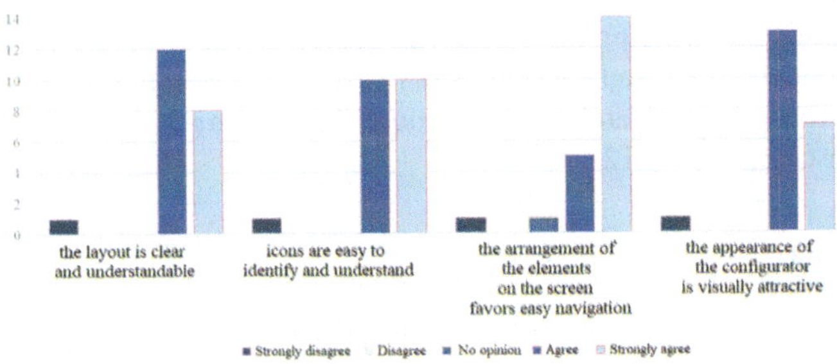

Fig. 10. The results of the user interface evaluation survey

Product Evaluation

Based on the questionnaires of the patients who received the orthoses for evaluation, it can be seen that most of the testers are either satisfied (2) or very satisfied (2). As for the compliance of the manufactured product with the expectations, almost all respondents were satisfied (3) or very satisfied (2) (Fig. 11).

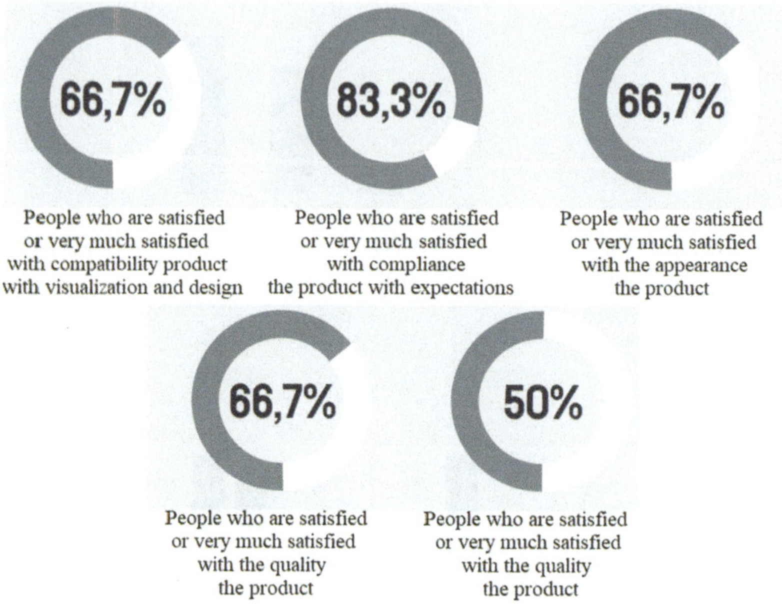

Fig. 11. Finished product evaluation

Overall System Assessment

All patients were satisfied with the services offered by the system. The majority of respondents (17) strongly believe that the system is helpful when purchasing a personalized orthopaedic device. According to almost all respondents (20), the main advantage of the system is the ability to personalize the product. Among the main advantages, most of the respondents mentioned the possibility of performing all the activities in one place - from 3D scanning to ordering the configured orthosis.

Each respondent expressed a willingness to use such a system should the need arise in the future. Test patients were asked to choose three main advantages system and the results are presented in Fig. 12.

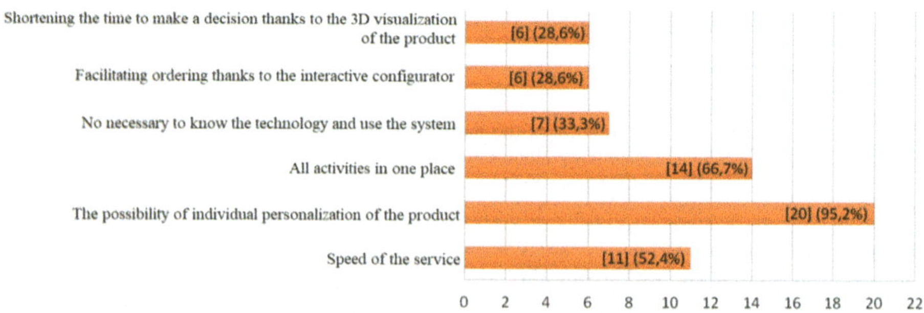

Fig. 12. Answer for question: What are the main advantages of the system in your opinion?

3.3 Discussion

The main objective of the study was to test and evaluate the performance of a prototype AutoMedPrint system for automated design and manufacturing of incremental orthopaedic and prosthetic supplies based on anthropometric measurements. Tests and questionnaires were carried out to evaluate the system with a group of test patients according to the planned scenario for the selected product - openwork wrist orthosis. The three main activities performed as part of the system operation, namely 3D scanning, design and rapid manufacturing, were tested.

Moving the scanner and starting the scanning depended on the operator controlling the movement of the device from the position of the laptop. This results in scan times that are operator dependent. Proper clothing for limb scanning is also important and should be communicated to the patient prior to the test. Wearing a blouse with or without short sleeves would improve patient comfort.

As a result of the evaluation of the scanning process, it was also suggested that some parts of the bench be repainted black, and that a black elastic thumb sleeve be worn during scanning. Black surfaces are not scanned, avoiding the need to manually clean scans and remove unnecessary artifacts. This approach has been already implemented and tested with success.

Errors in the CAD model and problems with generating data caused the authors to change the methodology of spline curve creation in the basic CAD model of the orthosis (original approach presented in [30]). The model was improved and in the tests conducted afterwards, orthoses were successfully generated (some with very minor errors) for all the patients.

The fabrication time mainly depended on the size of the orthosis as well as the printing strategy and orientation chosen by the patient. The orthoses for women were smaller than those for men, thus less material was used for them, making the printing time much shorter. In order to shorten the production time of the whole orthosis, both parts were printed in parallel on two machines.

The measured times show that the production time for one orthosis is not long and therefore future patients can receive the finished products the next day. Traditional methods of orthosis manufacturing do not offer this possibility.

The fit of the manufactured product was also controversial, as it can be concluded from the results that the orthoses did not fit every other person. For half of the patients (3), the orthosis proved too tight when attempted to be fitted. During the design process the offset from the limb was set at 1 mm. This value was verified in further studies on a group of several patients, and for adult patients was set at default of 3 mm, which was proven to fit all the patients, while not being too loose.

3.4 Conclusions

In conclusion, the results of the study showed that it is possible to obtain a fully functional and affordable orthosis in less than one working day utilizing an automated design process, allowing the patient to start the rehabilitation process immediately. Just one visit is enough to create a custom orthosis. In case of failure at any stage, it can be easily reproduced to get the correct result. This is a worldwide unique capability of the AutoMedPrint system.

Another advantage of the system is the possibility of operating in one location, as well as the lack of requirements for specialist knowledge of modern technologies on the part of both the operator and the patient. It would therefore be possible to implement such a system in medical facilities, without the need for skilled engineers or technicians. The patient has the possibility to actively personalize the product and the visualization of the product helps them to make decisions. Considering the results of the questionnaire completed by the study participants, patients are satisfied with the services offered by the system and express their willingness to use such a system if needed in the future.

The research results and conclusions were used in further stages of AutoMedPrint system testing. Further studies on the system are conducted with a younger group of patients, as well as individuals after a wrist joint injury requiring real treatment. This will help to discover other service and fitness issues and refine the system's performance.

References

1. Chua, C.K., Leong, K.F., Lim, C.S.: Rapid Prototyping: Principles and Applications, p. 420. World Scientific Publishing, Singapore (2003)
2. Syam, W.P., Mannan, M.A., Al-Ahmari, A.M.: Rapid prototyping and rapid manufacturing in medicine and dentistry. Virtual Phys. Prototyp. **6**, 79–109 (2011)
3. Banaszewski, J., et al.: 3D printed models in mandibular reconstruction with bony free flaps. J. Mater. Sci. Mater. Med. **29**, 23 (2018)
4. Kelly, S., Paterson, A., Bibb, R.J.: A review of wrist splint designs for additive manufacture. In Rapid Design. In: Proceedings of the Prototyping and Manufacture Conference, Loughbrough, UK, 15–16 December 2015, p. 12 (2015)
5. Gholamipour-Shirazi, A., Kamlow, M.A., T Norton, I., Mills, T.: How to formulate for structure and texture via medium of additive manufacturing-a review. Foods **9**, 497 (2020)
6. Velázquez, J.S., Cavas, F., Bolarín, J.M., Alió, J.L.: 3D printed personalized corneal models as a tool for improving patient's knowledge of an asymmetric disease. Symmetry **12**, 151 (2020)
7. Tanner, J.A., et al.: A three-dimensional print model of the pterygopalatine fossa significantly enhances the learning experience. Anat. Sci. Educ. **13**, 568–580 (2020)
8. Ivanova, O., Williams, C., Campbell, T.: Additive manufacturing (AM) and nanotechnology: promises and challenges. Rapid Prototyp. J. **19**, 353–364 (2013)
9. Novakova-Marcincinova, L., Novak-Marcincin, J., Barna, J., Torok, J.: Special materials used in FDM rapid prototyping technology application. In: Proceedings of the IEEE, 16th International Conference on Intelligent Engineering Systems (INES), Lisbon, Portugal, 13–15 June 2012, pp. 73–76 (2012)
10. Safka, J., Ackermann, M., Martis, D.: Chemical resistance of materials used in additive manufacturing. MM Sci. J. **6**, 1573–1578 (2016)
11. Dulchinos, J.: 3D Printing Trends: Five Major Developments (2019). https://www.jabil.com/blog/3d-printing-trends-show-positive-outlook.html. Accessed 17 Oct 2021
12. Tukuru, N., Gowda, K.P., Ahmed, S.M., Badami, S.: Rapid prototype technique in medical field. Int. J. PharmTech Res. **1**, 341–344 (2008)
13. Mavroidis, C., et al.: Patient specific ankle-foot orthoses using rapid prototyping. J. Neuroeng. Rehabil. **8**, 1 (2011)

14. Palousek, D., Rosicky, J., Koutny, D., Stoklásek, P., Navrat, T.: Pilot study of the wrist orthosis design process. Rapid Prototyp. J. **20**, 27–32 (2014)
15. Paterson, A.M., Bibb, R., Campbell, R.I., Bingham, G.: Comparing additive manufacturing technologies for customised wrist splints. Rapid Prototyp. J. **21**, 230–243 (2015)
16. Baronio, G., Volonghi, P., Signoroni, A.: Concept and design of a 3D printed support to assist hand scanning for the realization of customized orthosis. Appl. Bionics Biomech. **2017**, 8171520 (2017)
17. Andringa, A., van de Port, I., Meijer, J.W.: Long-term use of a static hand-wrist orthosis in chronic stroke patients: a pilot study. Stroke Res. Treat. **2013**, 546093 (2013)
18. Buonamici, F., et al.: A CAD-based procedure for designing 3D printable arm-wrist-hand cast. Comput. Aided Des. Appl. **16**, 25–34 (2019)
19. Aguado Pérez, J.: 3D Printed Orthosis Design; Polytechnic University of Valencia: Valencia, Spain (2018). http://hdl.handle.net/10251/104704. Accessed 17 Oct 2021
20. Huotilainen, E., et al.: Inaccuracies in additive manufactured medical skull models caused by the DICOM to STL conversion process. J. Cranio-Maxillo-Facial Surg. **42**, e259–e265 (2014)
21. Baronio, G., Harran, S., Signoroni, A.: A critical analysis of a hand orthosis reverse engineering and 3D printing process. Appl. Bionics Biomech. **2016**, 8347478 (2016)
22. Li, J., Tanaka, H.: Feasibility study applying a parametric model as the design generator for 3D–printed orthosis for fracture immobilization. 3D Print. Med. **4**, 1 (2018)
23. Lulkiewicz, M.: Evaluation of a prototype system of rapid design and manufacturing of orthopedic supplies. Master's thesis, Poznan University of Technology, Poznań (2021)
24. Górski, F., Wichniarek, R., Kuczko, W., Żukowska, M., Lulkiewicz, M.: Experimental studies on 3D printing of automatically designed customized wrist-hand orthoses. Materials **13**(18), 4091 (2020)

Concept of Individual Orthosis with 3D Pattern Structure for Upper Limb Using 3D Printing Technology

Agata Mrozek(✉), Ewa Tomaszewska, and Michał Rychlik

Institute of Applied Mechanics, Faculty of Mechanical Engineering, Poznan University of Technology, Jana Pawła II 24, 60-965 Poznań, Poland
agata.mrozek@doctorate.put.poznan.pl

Abstract. Reverse engineering and rapid prototyping found a wide range of applications in the area of orthopaedics. The development of these technologies enabled the manufacturing of orthoses and prostheses adapted individually to the patient's anatomy. Additionally, contemporary computer techniques allow to produce orthoses with various designs which help to keep the hygiene of limb, facilitate access to injury, and ensure ventilation of skin. The aim of this study was to develop the concept of individuated orthosis dedicated to the upper limb with pattern structure and removable elbow part to speed up the rehabilitation process. The designed orthosis consisted of 6 interconnected components. Additionally, instead of the standard separation edge along a straight line, a spiral curve was used. The proposed 3D pattern structure reduced mass by 31.3% compared to a full-volume component. The designed model of the orthopaedic appliance was subjected to strength analysis using the finite element method (FEM). Simulation results, in the form of stress and displacement distributions, were presented and discussed. Then, a physical model of the developed orthosis made using rapid prototyping technology (3D printing) was presented.

Keywords: Orthosis · 3D Printing · CAD · FEM simulation · Biomedical Engineering

1 Introduction

Over the last few years, many changes can be observed in the field of orthopaedics. This has been facilitated by development of reverse engineering techniques and rapid prototyping methods [1]. Additive manufacturing techniques enable to individualise orthopaedic appliances such as orthoses [2] and prostheses [3], increasing their comfort, functionality and adaptability to a specific kind of the patient injury.

One of the most commonly used methods of orthosis manufacturing is FDM (Fused Deposition Modeling) [4]. Pilot studies proved that application of FDM allows to obtain orthosis resistant to bending stress [5]. The mechanical properties of manufactured parts are affected by process parameters, for instance: layer height or infill density. These techniques provide the possibility to obtain low cost orthosis with low mass. Other

F. Gorski et al. (Eds.): MANUFACTURING 2022, LNME, pp. 16–30, 2022.
https://doi.org/10.1007/978-3-030-99769-4_2

commonly used additive manufacturing technology in orthopaedics is SLS (Selective Laser Sintering). Nevertheless, parts manufactured using these techniques are characterized by high porosity between fused powder particles. Moreover, the limitation of SLS application is high equipment cost [6].

A bone fracture occurs when the amount of applied load exceeds strength of the bone. The type of injury is influenced by factors like the direction and magnitude of the forces applied, the age and condition of the patient's skeletal system. The biomechanical stimuli influencing the optimal treatment process include proper alignment of the fragments and their stiffening, as well as proper compression [7].

In case of a bone fracture it is necessary to immobilise the bone. It not only ensures the correct structure union and optimal biomechanical conditions, but also prevents shock and secondary trauma. Depending on the type and location of the fracture, a variety of orthopaedic appliances are used [8, 9], in particular, to stiffen the injured body part. Moreover, for upper limb fractures it is important to use a lightweight appliances [10]. Any additional load adversely affects the injured part of the body. Long-term immobilisation of the upper limb affects the range of motion in the joint, especially the ability to straighten [11]. Therefore, the period of limb immobilisation should be as short as possible. Among the stabilizers of the upper limb, plaster casts, fiberglass casts, orthoses, and external stabilizers are distinguished [12].

In this study, having regard to the advantages of using additive manufacturing methods in orthopaedics [13, 14], the concept of orthosis with the ability to support a gradual increase in mobility of a joint was developed. The aim of the paper was to propose the approach that allows the creation of customised orthoses with their functionality adapted to the patient's anatomy and the stage of treatment of the upper limb. One of the main problems in respect to conventional plaster bandages is that the range of motion cannot be changed at a later stage of treatment. This often results in weakness of the extremity. Additionally, during the literature review it was noticed that existing solutions are mainly focused on orthosis involving only the forearm and wrist. Due to these facts, this study develops a concept that allows for a gradual increase in the mobility of the elbow joint. In order to create individualised orthopaedic appliances, basic geometry of the designed orthosis was created on the basis of upper limb anatomy. Additionally the new method of dividing the orthosis' elements was proposed. Instead of the standard separation edge along a straight line, a spiral curve was used. A 3D pattern structure has been proposed to reduce the weight of orthopaedic appliances.

2 Research Problem

2.1 Applied Materials for Orthosis Manufactured by 3D Printing

The applied material ought to be durable enough to ensure that the orthosis will fulfil its function during the entire period of treatment and rehabilitation. The recovery process may last several weeks therefore the material used should not cause skin allergies. An additional aspect that should be taken into account is the water resistance of the material. For the patient's comfort, it is important that the activities related to maintaining hygiene do not affect the ability of the orthosis to perform its function. One of the most essential requirements for the applied material is to provide the ability to achieve adequate load-bearing capacity at a relatively low mass.

Having regard to the presented requirements and the selected method of manufacturing (FDM), the materials to be used during the analysis were defined. The orthosis is assumed to be made of polylactic acid (PLA). This material fulfils the requirements for orthotics and also allows the use of additive manufacturing methods. PLA is obtained from materials of natural origin. It is a non-toxic, biodegradable material, which makes the material not only suitable for industrial use, but also for biomedical applications. PLA is classified as a thermoplastic material. The mechanical properties of PLA allow to maintain the functionality of the orthosis while minimizing its mass.

The material for the removable elbow part must be characterized primarily by its flexibility. Due to this fact, thermoplastic polyurethane (TPU) is used for the replaceable segment covering the elbow joint and the rings that connect the parts. The material for the removable part of the elbow is required to be flexible and be able to carry the load of this orthosis part. The materials properties are shown in Table 1. Presented characteristics were developed on the basis of materials specification sheets supplied by producers of the materials [15].

Table 1. Properties of materials used in the upper limb orthosis [18].

Property		PLA	TPU
Yield Strength	[MPa]	70	35
Young's Modulus	[GPa]	3.15	0.03
Poisson's ratio	[-]	0.36	0.48
Density	$[\frac{g}{cm^3}]$	1.24	1.27

2.2 Model of the Upper Limb Orthosis According to a New Concept

The advantage of using rapid prototyping techniques in orthotics is the possibility of individualizing orthopaedics appliances. The creation of the orthosis model begins with the acquisition of a 3D scan (by 3D structured light scanner) of the upper limb. In this study, male arm model from an open source database was used [16]. In order to apply the anatomical shape of the limb to create the orthosis model, the geometry had to be modified. Flexion of the elbow joint was performed using Rhinoceros 3D software. The orthosis' geometry was created on the basis of the obtained shape of the upper limb using Autodesk Meshmixer 3.5 and Inventor 2019 programs. The model thickness was 5 mm (Fig. 1a). It was assumed based on values recommended by other authors [1, 17].

In order to allow effortless assembly, orthosis is divided into several parts. Instead of the standard separation edge along a straight line, a spiral curve was used to reduce stresses and displacements between the elements. Application of spiral curve as detachment path determined location of separation edges. The line is designed in a manner that enables the possibility of safe (for patient) assembly process of components or part replacement. Connectors are located in easily accessible places. The orthosis model was divided into 6 parts (Fig. 1b). Two parts for the arm (part 1 and part 2), two parts for stabilising the elbow joint (part 3 and part 4) and two for stiffening the forearm (part 5 and part 6) were included. The part covering the elbow joint was designed in two variants. First variant is responsible for complete stiffening of the limb, whereas the second option is partially flexible to support the rehabilitation process. Component separation is made in order to enable the exchange of this part for a version made of more flexible material, which increases the range of motion of weakened structures in the area of the elbow joint.

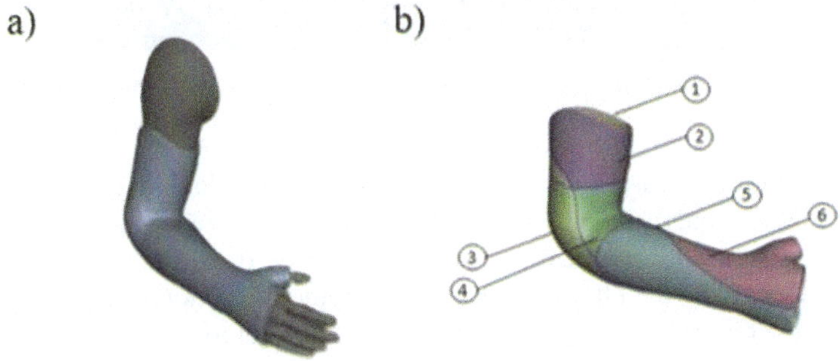

Fig. 1. Designed orthosis: a) preliminary model of the orthosis geometry obtained from 3D scanning, b) components of the orthosis with visible separation edges (1–2: parts covering the arm, 3–4: parts encompassing the elbow joint, 5–6: parts covering the forearm).

Due to the fact that the geometry of the orthosis is not monolithic and consists of several parts, connectors, which make it possible to match all components, were proposed. The designed joint consists of a flexible, durable ring, which is applied to two semicircles located on separate parts of the orthosis (Fig. 2). In order to verify the dimensions of connectors, a ¼-length section of the orthosis was designed. It was loaded with a force of 12 N (representing about ¼ of the upper limb mass). After conducting the simulation, it was found that the designed geometry of the connectors system fulfil the strength criteria for the selected materials. The maximum stress value was 5.1 MPa and occurred at the elastic ring (Fig. 3).

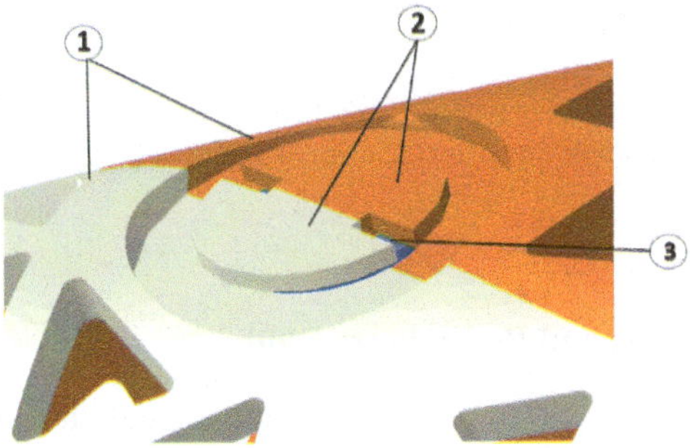

Fig. 2. Visualisation of the applied connector (1: combined elements, 2: semicircles as connector elements, 3: ring).

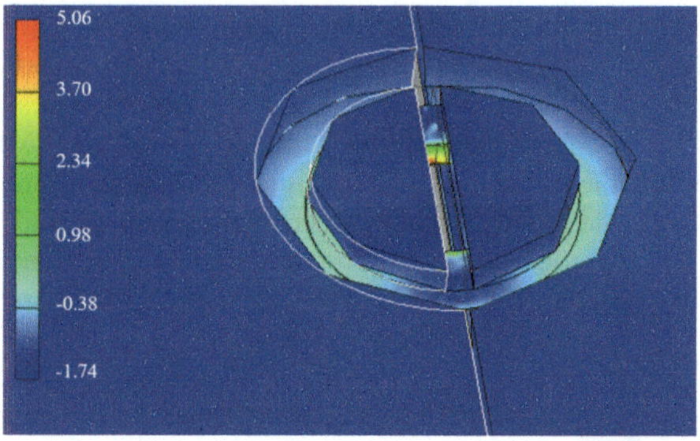

Fig. 3. The results obtained for connector- 1st principal stresses expressed in [MPa].

The assembly of the parts is uncomplicated. The oval and flexible geometry of the ring allows for proper shape matching of the orthosis to compensate for the increased volume of the limb due to swelling. The diameters of the elastic rings and the connectors were enlarged in the case of the elbow section, because their function is to connect the four components of the orthosis. After modelling the connectors, the orthosis model was divided into its final elements (Fig. 4). The development of the 3D pattern structure began with a preliminary analysis of the stress state in the orthosis, which was intended to identify the segments carrying the highest loads.

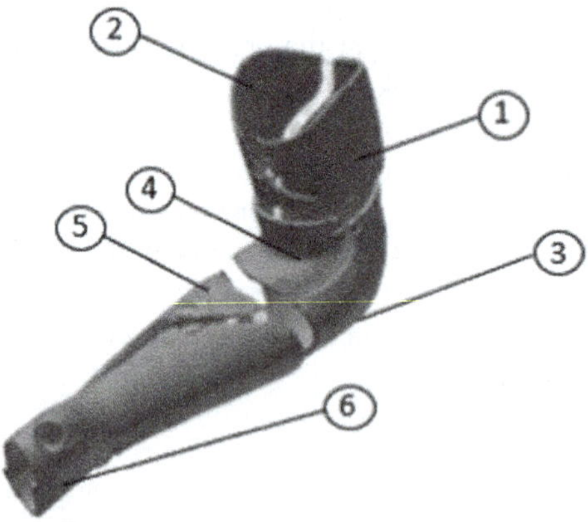

Fig. 4. Visualization of the orthosis with designed connecting elements after the final separation procedure of elements (1–2: parts covering the arm, 3–4: parts encompassing the elbow joint, 5–6: parts covering the forearm).

2.3 Loading Conditions and Preliminary Study

Performing an appropriate analysis using the finite element method (FEM) allows for verification of the orthosis behaviour during selected activities. The load conditions, which were taken into account in the selected analyses are shown in Table 2. The first analysis involved the case where the orthosis is loaded with the mass of the upper limb only. According to Gedliczka [18] and Buśkiewicz et al. [19], the mass of the adult upper limb is about 4.5 kg. In the second case of analysis, the model was loaded in the lower part of the orthosis, covering the wrist, with a force of 90 N corresponding to a possible impact or increased muscle force activity in the upper limb. The third simulation represents the load on the critical segment, from the point of durability, resulting from the work of the biceps brachii muscle during rehabilitation of the limb [18, 20]. A summary of inputs applied in the finite element analyses (FEA) was presented in Table 2. Static analyses were performed using finite element mesh which consist of tetrahedral elements (Table 3). All experiments were conducted using Autodesk Fusion 360.

Table 2. Applied loads and boundary conditions in the simulations: 1- fixed constraint, 2- load application surface.

Simulation	Applied loads and boundary conditions		Type of material
	Loading force	Boundary conditions	
Case I	45 N	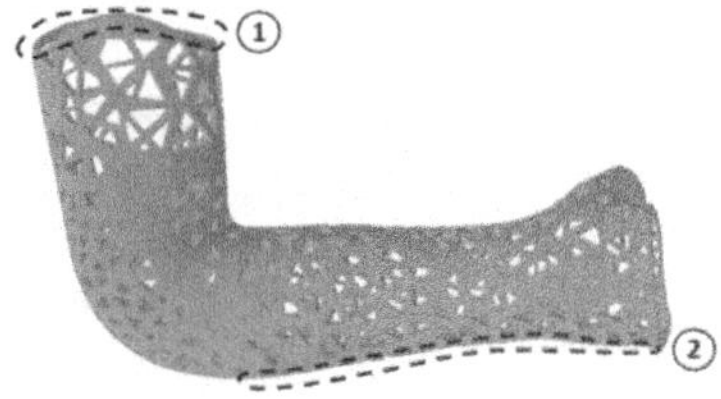	PLA
Case II	90 N	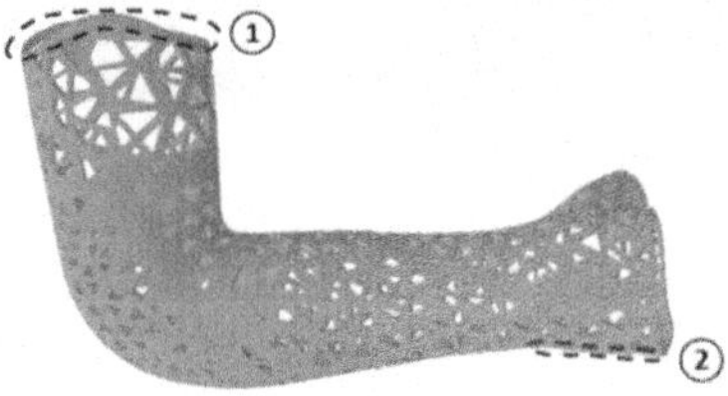	PLA
Case III	116 N	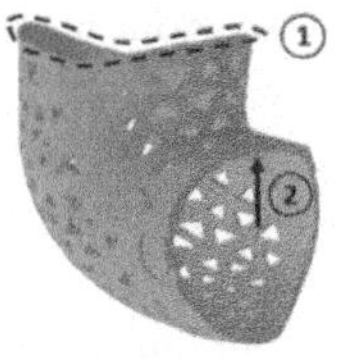	TPU

Preliminary analysis enabled to plan the initial topology of the pattern structure. The 3D designed pattern will allow for ventilation of the skin and increased access to the injured extremity. It could also be applied due aesthetic purposes. In this study it was

Table 3. Applied mesh parameters.

Parameter	Value
Number of nodes	929 447
Number of elements	552 002

assumed that the 3D pattern unit cells will be triangular shaped to simplify the process of controlling elements size.

The highest stress values were observed around the elbow joint. These stresses reach the value of 0.4 MPa (Fig. 5). This means that the mass of the orthosis could be reduced. In the arm and forearm sections, the percentage of mass reduction should be higher than in the area of the elbow joint. The segment with the maximum stress values is characterised by smaller dimensions of pattern elements (Fig. 6). For instance, the average length of the triangle edge in the part covering the arm was equal to 16 mm, while in the elbow part the average length of the pattern element was 8 mm. The proposed 3D pattern structure was made on the basis of sketches created in Autodesk Inventor 2019 software.

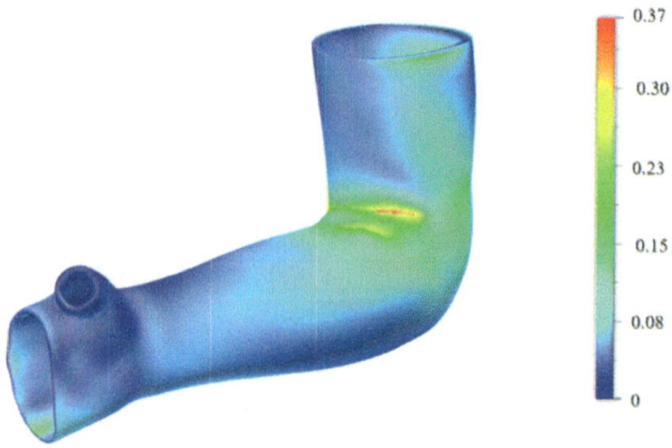

Fig. 5. Obtained von Mises stress distribution during preliminary study- stresses are expressed in [MPa].

The part, which encompasses the elbow joint is a replaceable element (Fig. 7). This modification was introduced in order to enable movement of the limb in joint during the advanced stage of treatment, when mobility of the limb is stimulated. To make it possible, a modification of the pattern structure within this component has been proposed. In the part involving the elbow joint, a notch was made near the cubital fossa, as shown in Fig. 8.

The elbow part would be manufactured (3D printed) from two materials with different levels of elasticity, according to the condition of the fracture and the doctor's recommendations. For example, material dedicated to the first phase of treatment would be made of a material with the lowest level of elasticity (in order to restrict the range of motion), and the second variant for rehabilitation purposes would be made of a material with higher elasticity (for increased range of motion during the final phase of treatment).

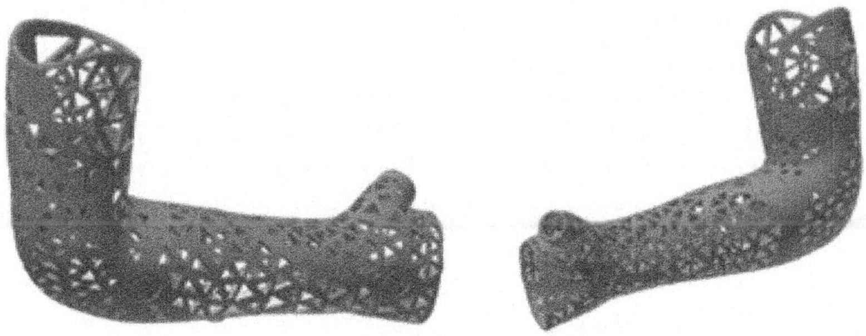

Fig. 6. Applied 3D pattern structure (triangle elements) – variable pattern elements size and distances between them depending on the level of stress in the defined area.

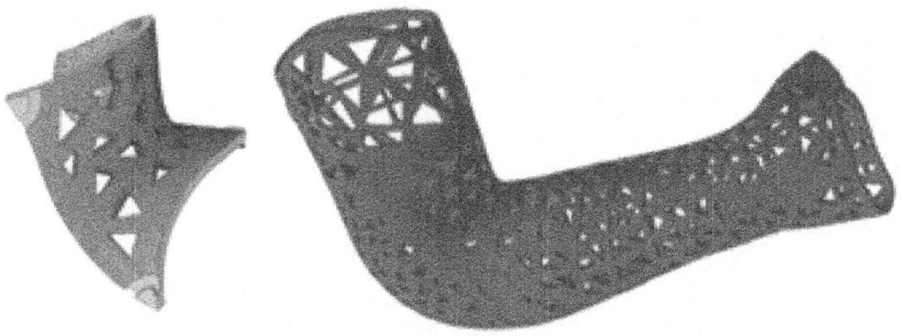

Fig. 7. Initial geometry of designed orthosis.

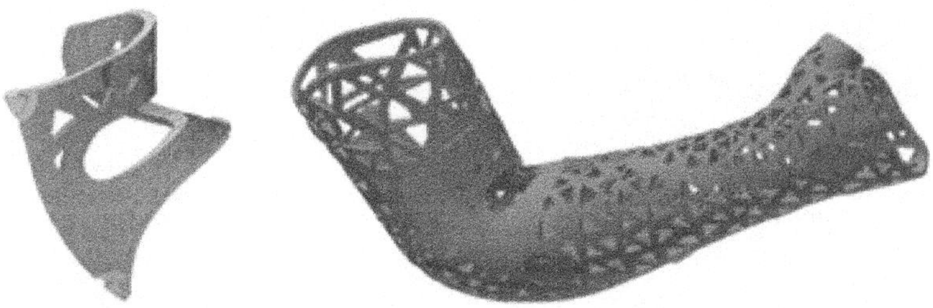

Fig. 8. Replaceable element in the elbow joint area- modification of the part covering the elbow joint.

3 Results of FEM Analysis of Orthosis Final Model with Pattern Structure

Every upper limb orthopaedic appliances are an additional load for extremity. Due to this fact, it is important to keep the mass of the orthosis as low as possible. Analysing the properties of the obtained structures, it was observed that the mass of proposed appliances decreased by 31.3% (Table 4). The distribution of pattern structure was planned on the basis of a preliminary strength analysis as presented in paragraph 2.3. The greatest decrease in mass was noted for the segment encompassing the patient's arm. In this part, the loss amounted to 42%.

As a next stage, an analysis of the designed orthosis with 3D pattern structure for different loading conditions was performed using FEM. Three static analyses have been conducted. The loads occurring during the use of the orthosis were simulated.

In the first simulation, a maximum stress of 9.5 MPa was recorded (Fig. 9). This stress concentration was observed in the part encompassing the elbow joint. Maximum displacement value was equal to 1 mm and it occurred in the part surround the patient's hand (Fig. 10).

Table 4. Orthosis mass comparison.

Orthosis model visualization	Mass of orthosis [g]	Percentage of mass reduction
Full volume structure 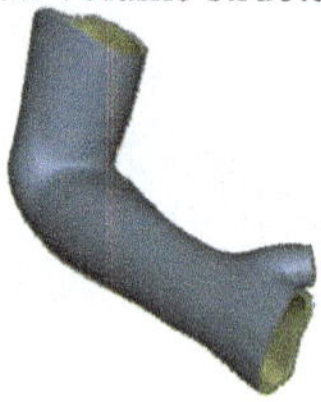	684	reference mass
Pattern structure and rigid elbow joint 	470	31.3 %
Pattern structure and flexible elbow joint 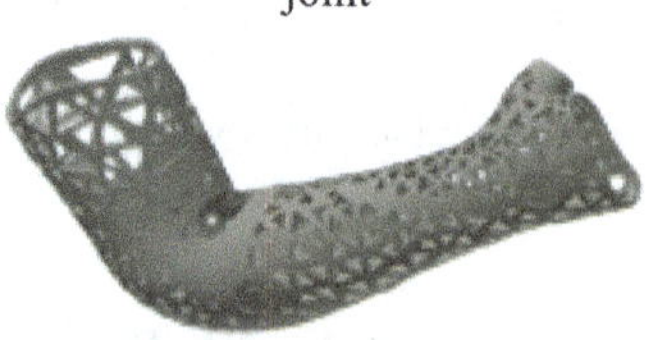	467	31.7 %

In case II, for a load of 90 N, a maximum stress value was observed in the interchangeable elbow part and it was equal to 29.75 MPa (Fig. 11). Maximum displacement value (3.4 mm) occurred in the lower surface of part encompassing the hand (Fig. 12).

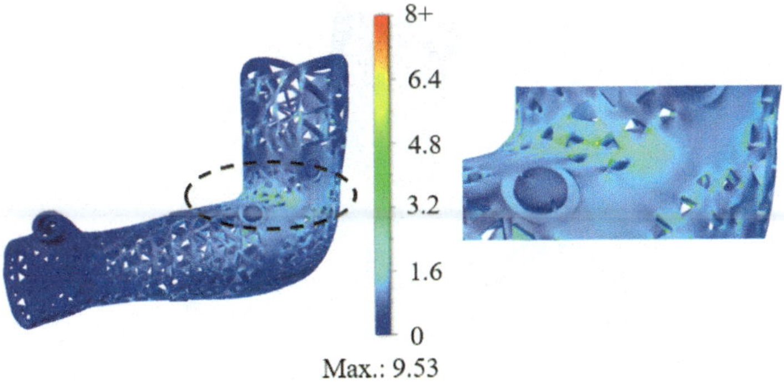

Fig. 9. The results obtained for the Case I - von Mises stresses expressed in [MPa].

Fig. 10. The results obtained for the Case I - displacements expressed in [mm].

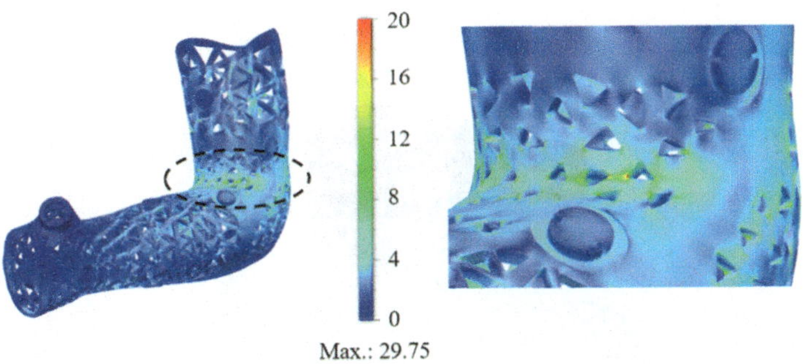

Fig. 11. The results obtained for the Case II - von Mises stresses expressed in [MPa].

Case III examined the durability of the critical orthosis part - the interchangeable elbow joint component. The material applied was TPU. The highest stress value was observed on surfaces adjacent to the notch made near to the cubital fossa (7.72 MPa) as shown in Fig. 13. The maximum displacement value was obtained on the lower surface of the component. It reached 31 mm (Fig. 14).

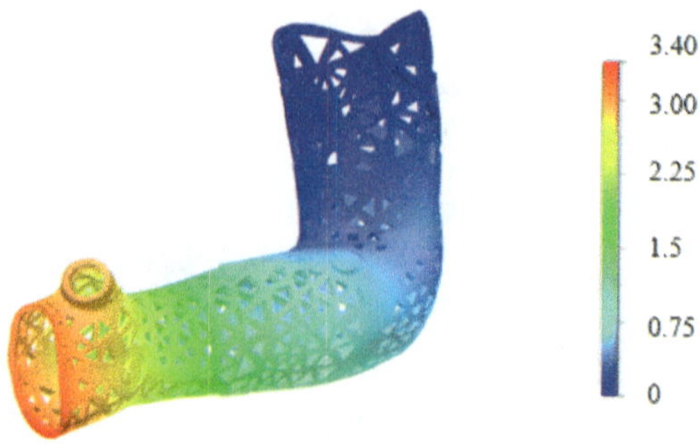

Fig. 12. The results obtained for the Case II - displacements expressed in [mm].

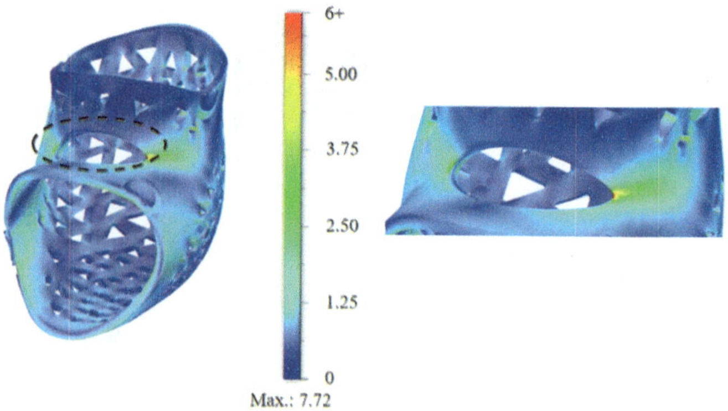

Fig. 13. The results obtained for the Case III - von Mises stresses expressed in [MPa].

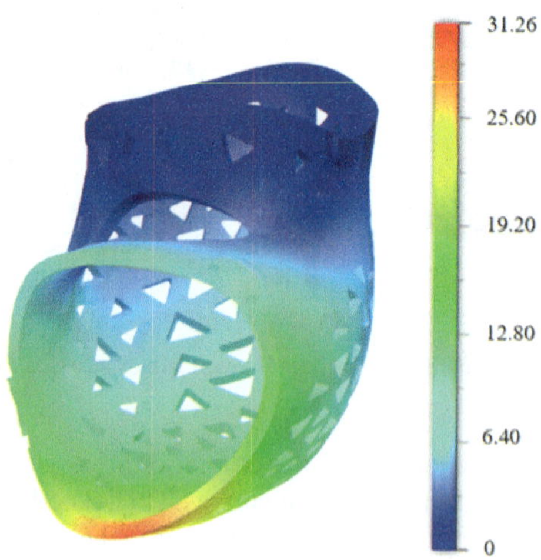

Fig. 14. The results obtained for the Case III - displacements expressed in [mm].

4 Manufacturing Process of Designed Orthosis

Due to the advantages of using additive manufacturing techniques in orthopaedics, the prototype of designed orthosis was made using the FDM method. The process parameters are shown in Table 5. Used 3D printer, which manufactured parts on the basis of the FDM method (Fig. 15), was XYZ da Vinci 1.0 Pro.

Table 5. Manufacturing process parameters.

Parameter	Value
Layer thickness [mm]	0.15
Part infill [%]	25-30
Support infill [%]	10
Extruder temperature [°C]	235-240
Bed temperature [°C]	90
Outline underspeed [$\frac{mm}{s}$]	15-26.25
Solid infill underspeed [$\frac{mm}{s}$]	17.50-18.75
Support structure underspeed [$\frac{mm}{s}$]	15-20

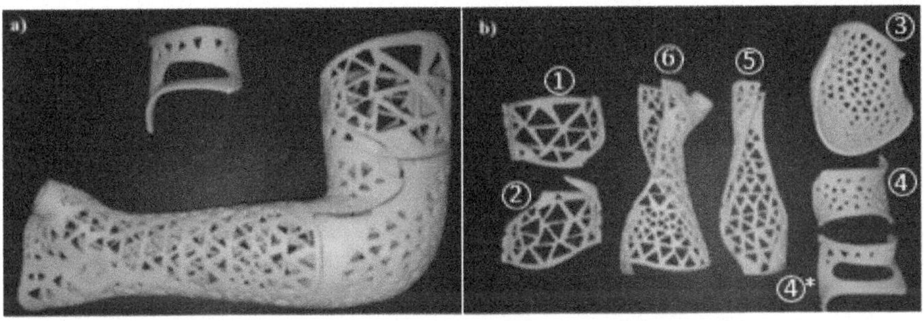

Fig. 15. Orthosis obtained in FDM process: a) view of the completed orthosis, b) view of individual parts (4*: second variant of replaceable elbow part).

5 Discussion

In all cases, the maximum stresses were within the safe range that did not exceed the yield strength of the selected materials. In case II, the maximum value of displacements is about three times higher compared to case I. The effect of orthosis individual components displacements on bone fusion should be verified. The analysis conducted for case III included a modified component at the area of the elbow joint. High displacement values were noted. Greater mobility within this component could provide an accelerated rehabilitation process with continuing stabilisation of the upper limb. The process of flexion and extension of the limb would be preserved in the 90° range. It should be highlighted that in order to obtain more accurate results verifying the influence of the applied loads, it would be necessary to conduct studies including the actual cooperation of the orthosis components.

The limb loading conditions applied in the analysis presented by authors are based on literature data and are estimates. In order to know the precise characteristics of forces acting on the orthosis, detailed research should be carried out.

The applied 3D pattern structure enables to reduce the mass of the orthosis. In addition, this solution allows for simpler maintenance of skin hygiene and ventilation. The assumed thickness of the orthosis was equal to 5 mm. The results of the FEA analysis indicate that there is a possibility to reduce this dimension.

Due to the possibility of customization of the orthosis model (Fig. 16), further research should be conducted to optimise the mass of the orthosis, not only with regard to the thickness of the orthosis or mass reduction in connection with a pattern structure, but also could be adjusted according to the age and weight of the patient.

The proposed geometry is manufacturable using additive methods. A test 3D print of the orthosis was made using the FDM method. The quality of printed parts and their mechanical properties depend on the parameters of the manufacturing process like infill density, extrusion temperature, infill patterns or layer thickness. The conducted analysis did not take into account the characteristics of the material which are influenced by the parameters of the manufacturing process. Further research should consider influence of manufacturing method and process parameters on obtained mechanical properties of the orthosis.

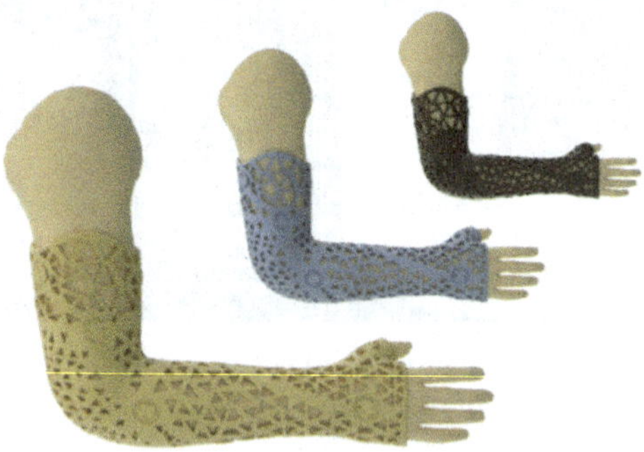

Fig. 16. Visualization of different colour variations of the proposed orthosis.

6 Conclusions

The conducted analyses showed that the orthosis fulfil functional requirements despite the 3D pattern structure. The advantage of this is that the mass of the orthosis was significantly reduced, which is beneficial from the point of view of the upper limb load. In addition, it increases the aesthetics and hygiene of the orthopaedic appliances.

In this paper, the approach that allows the creation of customised orthoses with their functionality adapted to the patient's anatomy and the stage of treatment of the upper limb was proposed.

The research is preliminary therefore parameters like the strength of the structure in relation to the printing parameter should be taken into account during the further analysis. The next step is to create a more accurate upper limb loading model and material properties (taking into account the heterogeneous structure of the material after printing). Additionally, having data from a detailed analysis of the stress distribution, research should be carried out to optimise the mass of the structure. Another approach to research may involve the analysis of the shape and size of the applied 3D structure pattern, which may occur in many different variations.

Acknowledgement. The possibility of printing was provided by the Department of Virtual Engineering of Poznan University of Technology, IWP Scientific Association "STANTON" and Konrad Łyduch MSc. Eng. for whom particular thanks are expressed.

The presented research results were funded with the grant 0612/SBAD/3576 allocated by the Ministry of Science and Higher Education in Poland.

References

1. Baronio, G., Volonghi, P., Signoroni, A.: Concept and design of a 3D printed support to assist hand scanning for the realization of customized orthosis. Appl. Bionics Biomech. **2017**, 8171520 (2017). https://doi.org/10.1155/2017/8171520
2. Faustini, M.C., et al.: Manufacture of passive dynamic ankle-foot orthoses using selective laser sintering. IEEE Trans. Biomed. Eng. **55**, 784–790 (2008). https://doi.org/10.1109/TBME.2007.912638
3. Hao, Y., et al.: 3D printing hip prostheses offer accurate reconstruction, stable fixation, and functional recovery for revision total hip arthroplasty with complex acetabular bone defect. Engineering **6**(11), 1285–1290 (2020). https://doi.org/10.1016/j.eng.2020.04.013
4. Oud, T.A.M., Lazzari, E., Gijsbers, H.J.H., Gobbo, M., Nollet, F., Brehm, M.A.: Effectiveness of 3D-printed orthoses for traumatic and chronic hand conditions: a scoping review. PLoS ONE **16**(11), 1–10 (2021). https://doi.org/10.1371/journal.pone.0260271
5. Palousek, D., Rosicky, J., Koutny, D., Stoklásek, P., Navrat, T.: Pilot study of the wrist orthosis design process. Rapid Prototyp. J. **20**(1), 27–32 (2014). https://doi.org/10.1108/RPJ-03-2012-0027
6. Kumar, R., Sarangi, S.K.: Design, applications, and challenges of 3D-printed custom orthotics aids: a review. In: Pratap Singh, R., Tyagi, M., Panchal, D., Davim, J.P. (eds.) Proceedings of the International Conference on Industrial and Manufacturing Systems (CIMS-2020). LNMIE, pp. 313–328. Springer, Cham (2022). https://doi.org/10.1007/978-3-030-73495-4_22
7. Błaszczyk, J.W.: Clinical Biomechanics. Textbook for medical and physiotherapy students. 1st edn. PZWL, Warsaw (2004)
8. Einhorn, T.: A Enhancement of fracture-healing. J. Bone Joint Surg. **77**(6), 940–956 (1995). https://doi.org/10.2106/00004623-199506000-00016
9. Einhorn, T.A., Gerstenfeld, L.C.: Fracture healing: mechanisms and interventions. Nat. Rev. Rheumatol. **11**(1), 45–54 (2014). https://doi.org/10.1038/nrrheum.2014.164
10. Fromme, N.P., et al.: Design of a lightweight passive orthosis for tremor suppression. J. NeuroEng. Rehabil. **17**(47), 1–15 (2020). https://doi.org/10.1186/s12984-020-00673-7
11. Mehlhoff, T.L., Noble, P.C., Bennett, J.B., Tullos, H.S.: Simple dislocation of the elbow in the adult. Results after closed treatment. J. Bone Joint Surg. Am. **70**(2), 244–249 (1988)

12. Thompson, S.R., Zlotolow, D.A.: Handbook of Splinting and Casting, 1st edn. Elsevier, Philadelphia (2012)
13. Kim, H., Jeong, S.: Case study: hybrid model for the customized wrist orthosis using 3D printing. J. Mech. Sci. Technol. **29**(12), 5151–5156 (2015)
14. Cha, Y.H., et al.: Ankle-foot orthosis made by 3D printing technique and automated design software. Appl. Bionics Biomech. **2017**, 1–2 (2017). https://doi.org/10.1155/2017/9610468
15. http://www.matweb.com/. Accessed Jan 2019
16. https://www.3dcadbrowser.com/3d-model/arm-male. Accessed Jan 2019
17. Li, J., Tanaka, H.: Feasibility study applying a parametric model as the design generator for 3D–printed orthosis for fracture immobilization. 3D Printing Med. **4**(1), 1–15 (2018)
18. Gedliczka, A.: Human Measurement Atlas - Data for Ergonomic Design and Assessment. CIOP, Warsaw (2001)
19. Buśkiewicz, J., Grabski, J.K., Walczak, T.: Guide to Laboratory Exercises in Engineering Biomechanics, 1st edn. PUT Publishing House, Poznań (2015)
20. Shahid, M.K., Fletcher, M., Robati, S., Pemmaraju, G.: The biomechanical forces that act on the elbow joint. EC Orthop. **1**(1), 1–11 (2015)

Design and Additive Manufacturing of an Individualized Specialized Leg Orthosis

Filip Górski[1(✉)], Justyna Rybarczyk[1], Przemysław Zawadzki[1], Wiesław Kuczko[1], Natalia Wierzbicka[1], Magdalena Żukowska[1], and Sabina Siwiec[2]

[1] Faculty of Mechanical Engineering, Poznan University of Technology, Piotrowo 3, 60-965 Poznań, Poland
filip.gorski@put.poznan.pl

[2] Faculty of Health Sciences, Poznan University of Medical Sciences, 70 Bukowska Street, 60-812 Poznań, Poland

Abstract. The paper presents studies on design and additive manufacturing of individualized ankle foot orthoses (AFOs) for a patient with a spina bifida. Main aim of the presented research and practical work was to design a functioning, stable and usable pair of 3D printed leg orthoses, which could be used for walking, as well as swimming and performing various activities in the water. The customizable orthoses were designed using a prototype of AutoMedPrint system developed by the authors. Then they were 3D printed using FDM technology, of various materials. The processes of 3D scanning, design and manufacturing, as well as testing with the patient are presented. In the end, an usable orthosis was obtained, which will be further processed and its design automated.

Keywords: Medical 3D printing · Additive manufacturing · Individualized orthoses · Customization · 3D scanning

1 Introduction

Orthoses are frequently used contraptions which help to maintain constant position of selected body parts of a given patient and protect these parts from external or internal harm from unwanted loading or movement. They are used in short-term healing when dealing with injuries or for long-term treatment of common conditions, such as cerebral palsy. Usually their purpose is achieved by building a shell, protecting body tissues around a selected joint from damage and immobilizing these joints. In the case of specific requirements, orthoses are also applied to make certain joints achieve a determined angular position, necessary during healing or treatment of long-term illnesses [1]. Generally, two options are available for the patients. The most popular, cheapest solutions are mass-produced and made in a variety of sizes. In terms of use comfort and healing properties, it is much more advisable to use individualized orthoses, made for one, specific person and theirs measured anatomical features [2].

F. Gorski et al. (Eds.): MANUFACTURING 2022, LNME, pp. 31–44, 2022.
https://doi.org/10.1007/978-3-030-99769-4_3

Orthoses and prostheses these days are often produced using additive, or layered manufacturing processes [3–5], known popularly as 3D printing. These processes are especially well suited for creation of customized, organic shapes, as no tooling is required. The most popular process of Fused Filament Fabrication (FFF), known also under the name of Fused Deposition Modeling (FDM) is low cost both in terms of machines and thermoplastic materials available. Therefore, it is very popular in medicine [6–8].

Additive manufacturing is considered as a modern approach to manufacturing orthopaedic supplies. A typical, traditional process involves manual activities, such as measuring a patient by making a "negative" using a plaster cast, then making a positive model. Then, lamination is performed, manually, yielding composite orthoses, for example based on glass fiber [9]. This process is used by many manufacturers, also by the biggest producers. The traditional processes are not repeatable, because they rely on manual skills, and the patient usually has to wait quite long for a finished orthosis. The obtained laminate is also difficult to recycle. Apart from this, there is no digital storing of patient data, for possible re-iteration of design.

The modern process introduces repeatability, as patient's anatomy is digitized and stored, by means of 3D scanning, usually. After scanning data processing, usually a work of a biomedical engineer is required to design an individualized orthosis, fulfilling anatomical (medical) and technical requirements. Then the orthosis is manufactured, often by industrial 3D printing.

One of the largest problems in 3D printing of individualized orthoses is requirement of specialized engineering knowledge. The anatomical measurement of a given person (the patient) must be first gathered and then processed to obtain usable data. This can generate a lot of inaccuracies [10, 11]. Obtaining a shape requires many hours of advanced modelling in CAD systems. Additionally, 3D printing of thermoplastic materials with satisfying values of accuracy and strength is difficult [12, 13]. There are constant studies on how to make the data gathering, processing and manufacturing easier and more available in general medical practice. Automation of certain engineering tasks seems a promising direction [14–16].

The modern process of orthoses generation is more robust, at the same time generating less waste, but it takes time and needs to be performed by a skilled engineer. This makes it much less available for large groups of patients than the traditional process and time of delivery can be also very long, although shorter than in the traditional process [10]. However, for a single, specific patient, the digital process makes it easier to quickly create and test many iterations, of which a final one can be subjected to design automation. Such a process is presented in this paper, demonstrating how modern process can lead to obtaining a usable, cheap, 3D printed orthosis for a specific patient.

2 Research Methodology

2.1 Aim of the Work

The studies presented in the paper are a case study performed in cooperation with a single patient. Main aim of the presented research and practical work was to design a functioning, stable and usable pair of 3D printed leg orthoses, which could be used for walking, as well as swimming and performing various activities in the water. The aim was to utilize the automation available in the previously designed AutoMedPrint system and rapidly create a set of iterations of the orthoses on the basis of 3D scanning of patient's leg. Three various designs and two different materials were iteratively tested, to finally come up with a solution both fitting the patient and enabling him to walk and perform all the necessary activities.

2.2 AutoMedPrint System

The research was realized in the scope of the project "Automation of design and rapid manufacturing of individualized orthopedic and prosthetic supplies on the basis of data of anthropometric measurement". The AutoMedPrint (AMP) is a system in development by the authors. It has a task of automated design and production preparation of individualized orthopedic supplies—mainly limb orthoses and upper limb prostheses. The system's concept is described in the authors' earlier works [17]. The principle of operation of the AMP system is shown in Fig. 1.

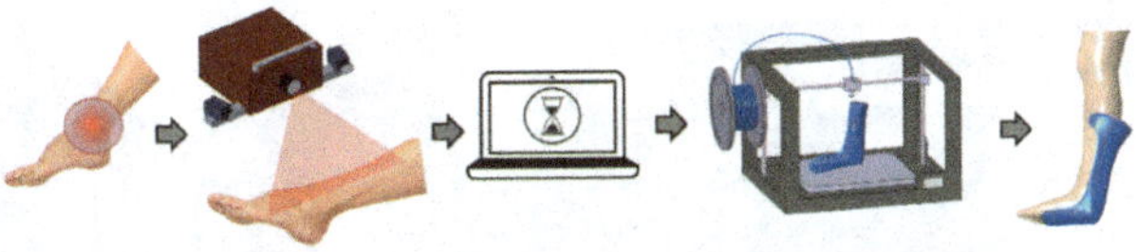

Fig. 1. The AutoMedPrint system – main idea [18]

The main stages of using the system are as following:

1. 3D scanning of patient limb, automated data processing,
2. automated CAD design of a selected product, based on 3D scanning data,
3. semi-automated preparation of 3D printing process and its realization,
4. fitting and testing with the patient.

Previous successful use cases of the system include wrist hand orthoses [17] and prostheses [19]. The system has been already used to produce AFOs for younger children and attempts were made at making leg stabilizer for adult patients. However, the specialized type of orthosis presented in the paper poses a significant challenge and thus it was introduced to the system as an entirely new type of product.

2.3 Patient Data and 3D Scanning Process

The patient is a 13-year old male volunteer, who was diagnosed with a spina bifida (section L4) and lower limb paralysis. In 26th week of pregnancy, as a fetus, was subjected to intrauterine surgery and as 12-year-old, right hip was surgically reconstructed. As a results of surgeries and intense rehabilitation since birth, the patient is able to walk despite lack of sensory input from lower half of his body. However, he is unable to walk without dedicated, specialized, very expensive orthoses, that have to be frequently replaced due to body growth. What is more, side effect of the hip surgery is shortening of the right leg by several centimeters, which requires the orthoses to be of asymmetric construction. Currently used orthoses are unfit for the water activities and patient is unable to walk to and from the water bodies (e.g. pool, lake), which severely affect his independence, comfort and set of desired activities. There are currently no orthoses commercially available for such patients, with intention of use in the water.

The patient was 3D scanned using the AutoMedPrint system hardware and automated software. Figure 2 presents the scanning process. Both legs were scanned in the "right angle" sitting position (knee bent at the right angle, parallel foot and femur) and straightened, laying position. The patient was scanned using two scanners: David SLS-3, installed on the automated work stand and EinScan Pro – a manual, handheld 3D scanner. During the scanning, a physiotherapist was present to select the most comfortable and anatomically correct positions.

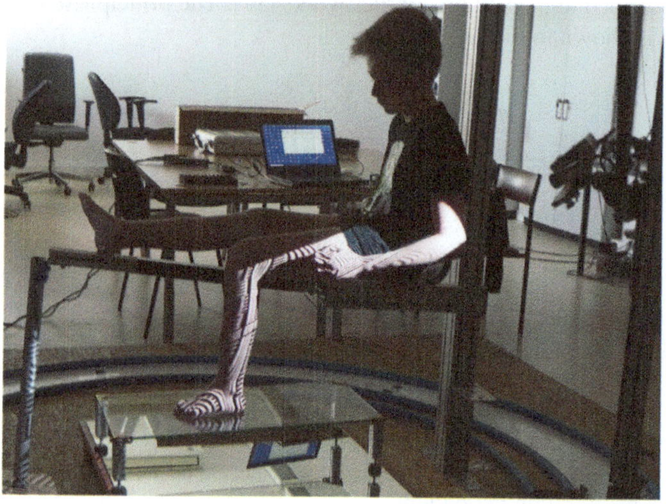

Fig. 2. 3D scanning of patient lower limbs using the AutoMedPrint system

Complete work with the patient (scanning of both legs in two positions using two scanners) took approximately 45 min. Another 30 min were spent for the joining, cleaning and reconstruction of the scans. The manual correction of the reconstructions took approx. 3 h of work of biomedical engineer and a physiotherapist.

The scans were subsequently joined, cleaned and limb model was reconstructed of them, mostly using automated algorithms embedded inside the AutoMedPrint system. The algorithms are based on the MeshLab open source software (version 2019). As an end result, 3D reconstructions in form of STL file – triangular mesh – were obtained for each leg, position and scanner (total of 8 limb models), example of such mesh is shown in Fig. 3. The meshes were then manually corrected using the GOM Inspect software, according to physiotherapist recommendations. The corrected meshes were used for the design stage.

2.4 Orthosis Design

The first step of design was getting coordinates of selected points directly from the reconstructed mesh, using scripts (recorded in the MeshLab software and then modified for purpose of automation). This process involved filtering and conversion, to obtain an Excel spreadsheet containing solely the filtered point coordinates. The point coordinates were used to construct the sets of splines, which created the basic shape of the orthosis, using the 3D Experience modelling software (CATIA v6).

Guidelines for the preparation of the initial project were as following:

- orthosis intended for contact with water, adding holes on the surface, allowing free flow of water and reducing the mass of the orthosis;
- required close adjustment to the anatomical shape of the patient's legs, ensuring stability during movement and comfort of use in water;
- wall thickness of the orthosis not exceeding 4 mm;
- the inner layer of the orthosis is lined with a textile material that separates the skin from the material of the orthosis;
- multi-part structure of the orthosis (several independent bodies), parts connected in a detachable manner (except for the straps, which are to be mounted with rivets);
- general division of the orthosis structure divided into two areas: upper (calf) and lower (foot);
- upper and lower parts are detachably connected with each other by means of a connector fixed with screws;
- upper part to stabilize the calf (adding tightening tapes or a mechanical lock);
- lower part ensuring the stabilization of the foot (addition of tightening straps or a mechanical lock);
- in the orthosis for the right leg, it is necessary to add a special support for the foot, ensuring the leveling of anatomical differences of the patient;

The initial model was created as a set of splines based on the original 3D scan of the lower limb. The splines were used to create surfaces, that were later converted into a solid (Fig. 3).

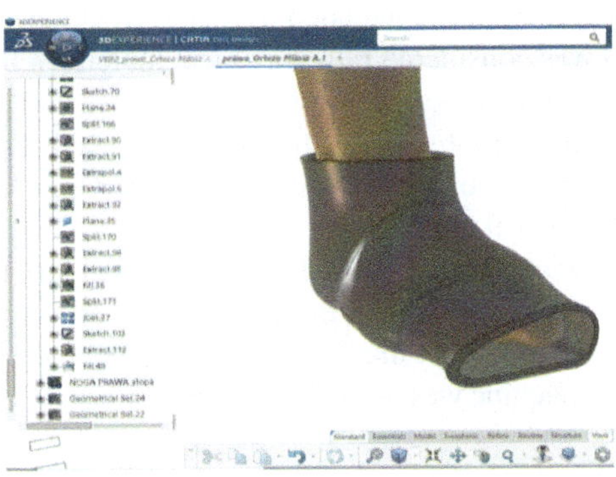

Fig. 3. 3D modelling of the orthosis

After the manufacturing and testing of the first design, another design was created. This process was repeated 2 more times – 3 subsequent design versions were obtained. Over the course of changing from version 1 to 2, design methodology was altered for easier change introduction. A wireframe model was introduced (Fig. 4), to make it easier during the later design automation. The basis for building the model were control points in specific sketches, connected with each other by lines (straight and / or curves). The logic of dependence of individual sketches of the skeleton was proposed, adopting specific hierarchical rules (superior and dependent points). This approach allowed for dynamic changes to be made while the modeling was in progress and was to guarantee the possibility of introducing changes in the future. On the basis of the prepared skeleton model, the surfaces of individual elements of the upper and lower parts were created.

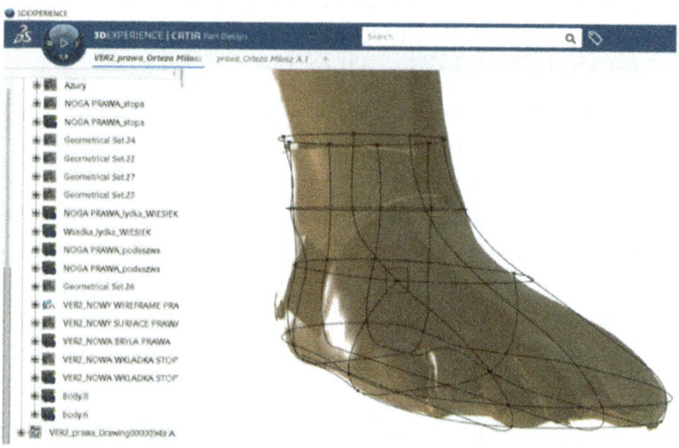

Fig. 4. Second iteration – wireframe model

2.5 Manufacturing

Each iteration of the product was manufactured additively, using Fused Deposition Modelling technology. Two materials were used as a building material. For the first, test iteration, PET-G material was used for the main orthosis body and the connecting beam was 3D printed of nylon. For the other two iterations, two materials were used: PET-G and nylon (PA-12) and the final material intended for use is nylon, due to its better mechanical properties (PET-G was considered rather as a test material). In the second iteration, the connecting beams were made of carbon fiber – epoxy resin laminate (in both material versions). Minimal amount of support structures was used. The layer thickness was 0,35 mm, infill was 60% and the build orientation was vertical (leg axis approximately parallel to the machine vertical axis).

For PET-G, the machine used was of Delta type – all the movement is realized by the extrusion head located at the end effector of the vertical, inverted Delta robot. The TEVO Little Monster low-cost machine was used for that purpose. For the nylon material, Zortrax M300 machine was used. The program was prepared using Simplify3D software.

After the 3D printing, the orthoses were subjected to basic post processing (support removal, grinding and removing sharp leftovers) and then were lined up inside using the EVA foam of various thickness (2–4 mm). Thicknesses and locations of the EVA foam were directly shown by the physiotherapist. The foam was joined to the orthosis inner surfaces using a waterproof double-sided adhesive tape. For the two first design iterations, the elements were joined using nuts and bolts.

2.6 Verification and Testing

The test procedure consisted of the following experiments:

1. Visual testing and general assessment.
2. Fitting test – patient in sitting position.
3. Standing test – with and without support.
4. Walking test – with and without support.

All experiments and assessments were performed by a team consisting of biomedical engineers and a physiotherapist. Tests number 2, 3 and 4 were conducted subsequently – after obtaining comfortable fitting, the patient tried to stand, first with assistance of physiotherapist, then without support. After successful standing, next step was walking, first several steps with assistance, then a longer distance (~10–15 m back and forth) with assistance and then an attempt without any assistance. For each orthosis, the stage of testing reached was noted, along with the observations and comparisons with the currently used commercial orthosis.

3 Results

3.1 Design Results

Differences between the subsequent designs are shown in the Fig. 5 (right leg only). The first iteration had the highest count of 3D printed parts – it contained the calf, the foot, the knee support, foot cover and connecting beam, plus an additional footing for the shorter leg. In the second iteration, the 3D printed beam was ditched in favor of a carbon fiber laminate and the foot cover was also removed. In the third iteration, calf and foot were merged together for better stability.

3.2 Manufacturing Results

All the iterations were manufactured successfully. No errors were recorded and processes were successfully realized. Times and material consumptions are shown in Table 1. Each single orthosis required approximately 2–3 h of manual work. It consisted of approx. 15 min of support removal and cleaning, above 1 h of foam cutting and gluing for the calf part and below 1 h for the foot part. Figures 6–8 present the finished orthoses (all three iterations discussed in this paper).

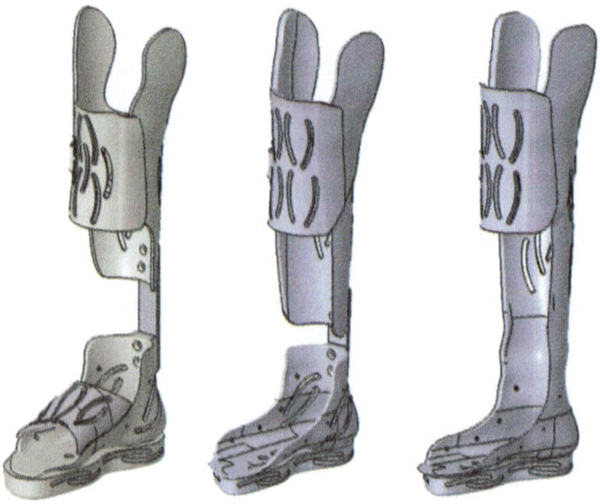

Fig. 5. Three iterations of design (1–3 left to right)

Table 1. Time and material consumption for all variants of the orthoses

Version	Side	Material	Time [min]	Material [g]	Support [g]
V1	Right	PET-G	1757	717	18.5
	Left		1141	495	13.5
V2	Right	PET-G	1391	799	20.5
	Right	nylon	4026	431	23
	Left	PET-G	1262	593.5	23.5
	Left	nylon	3947	414	25
V3	Right	PET-G	1484	812.5	8
	Right	nylon	3915	567	5
	Left	PET-G	1441	681.5	32
	Left	nylon	3332	432	7

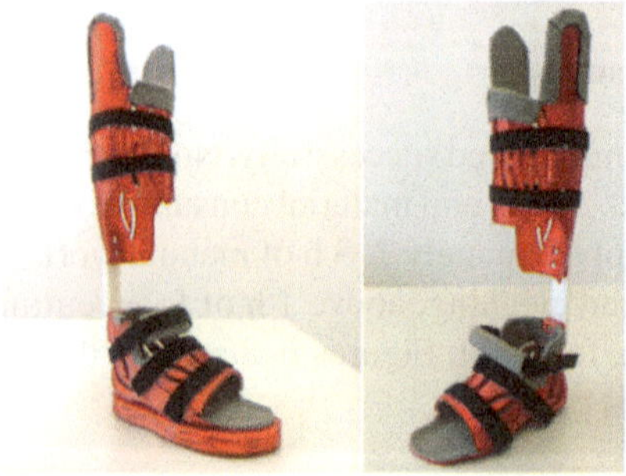

Fig. 6. The manufactured orthosis, design 1 – PET-G

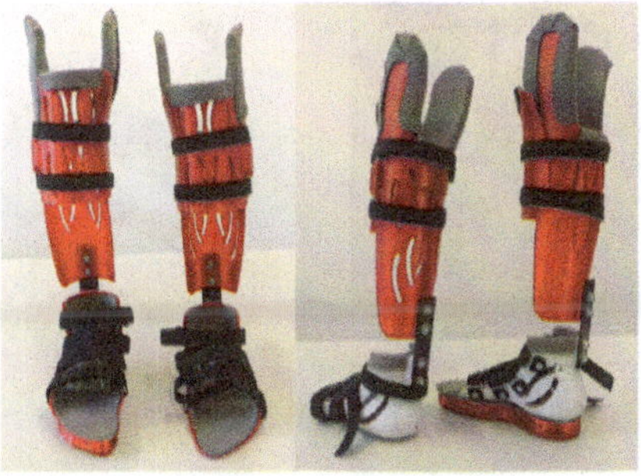

Fig. 7. The manufactured orthosis, design 2 – PET-G and nylon

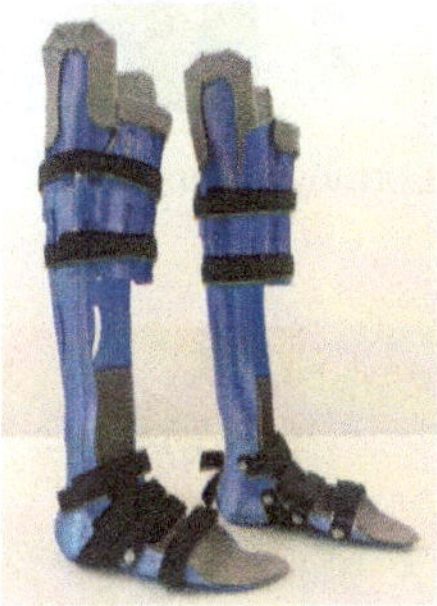

Fig. 8. The manufactured orthosis, design 3 – PET-G

3.3 Testing Results and Discussion

In general, all the iterations made it possible to reach the last stage (walking), although only the last iteration allowed somehow independent walking (with minimum to no support from other persons).

The main observations were divided into subsequent orthoses – pairings of design iteration and specific material. As every single orthosis passed through the visual control and general initial assessment, this stage will be omitted.

1. Two-piece nylon beam orthosis, PET-G material

1) fitting test (Fig. 9) – the orthosis was 6–7 cm too long and 5 cm too wide in the midfoot, too narrow in the ankle area by 2–3 cm, the heel did not reach the inner surface; it was also 8–9 cm too wide under the knee and the support straps broke off in one place; the additional piece securing the foot from upwards was found as unnecessary and thus removed from subsequent iterations;
2) standing test (Fig. 10) – no possibility of self-supporting, only with the help of other persons or a wall, patient swaying forward and backward (nylon beam was too elastic);
3) walking test - no possibility of independent walking, possibility of walking by holding to a wall or with the help of other persons, very low stabilization in knee area, the patient was leaning forward and to the right side.

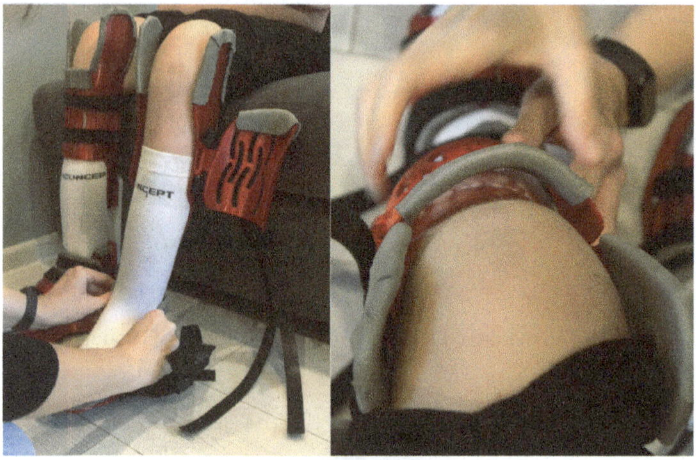

Fig. 9. Fitting of orthosis, version 1 – visible problems

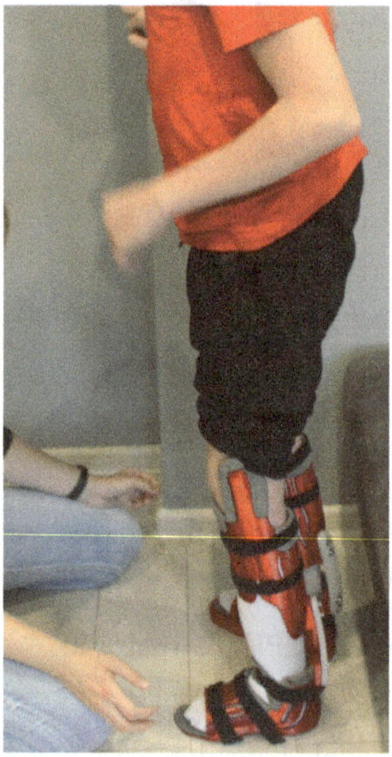

Fig. 10. Standing test of orthosis, version 1, visible bending of the nylon beam

2a. Two-piece carbon fiber beam orthosis, PET-G material

1) fitting test – no major problems, slightly too loose in the knee area;
2) standing test (Fig. 11a)- the patient was able to support himself, but due to slight rocking forward and backward, a bit of support from other persons was needed;

3) walking test - the orthosis broke after a few steps in the foot part, in the beam assembly area (Fig. 11b).

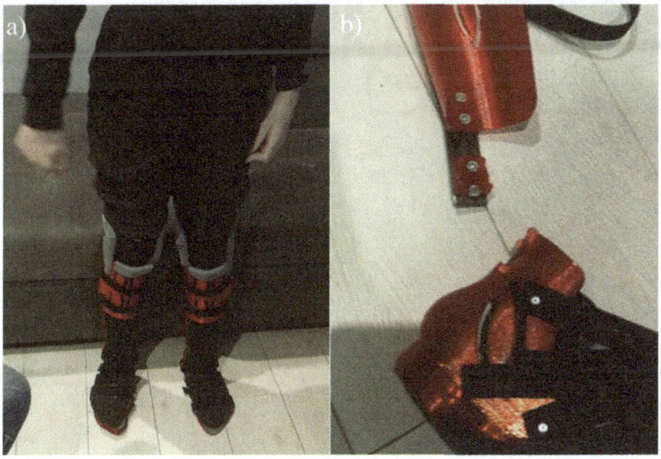

Fig. 11. Standing and walking test of orthosis, version 2, a) self-supported standing, b) broken orthosis with visible failure location

2b. Two-piece carbon fiber beam orthosis, nylon material

1) fitting test – the same as PET-G orthosis;
2) standing test - the same as PET-G orthosis;
3) walking test – assisted walking (with help of other persons) was possible, but at first attempt at independent walking, the orthosis broke, in a similar area as the PET-G orthosis.

3. One-piece, no beam, PET-G material

1) fitting test – no problems;
2) standing test – the patient was supporting himself, but due to the leaning forward, slight protection was needed;
3) walking test – assisted walking was fine (Fig. 12); the patient was able to take a few steps by himself, but bending and twisting his knees towards the outside. With the protection of third parties, the patient was able to walk a distance of approx. 25 m, but the knee twisting to the left side was noticeable. Walking was much more stable in the shoes than without them.

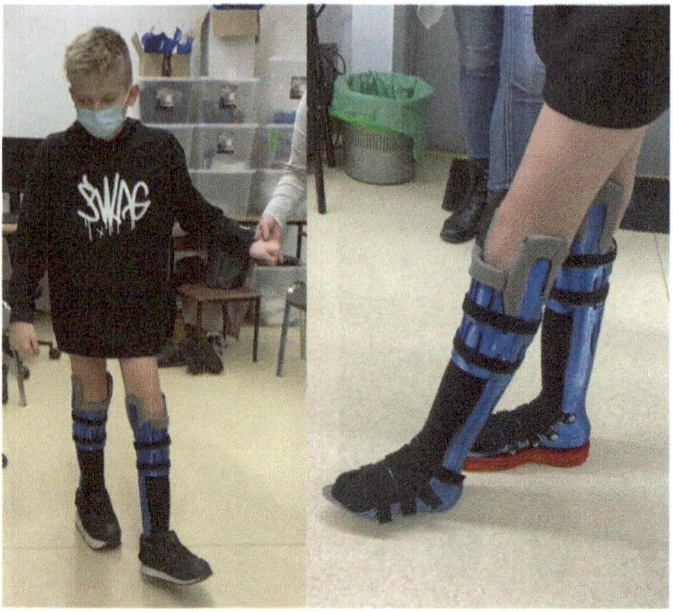

Fig. 12. Walking test of orthosis, version 3, with and without shoes

The main conclusions of all tests are as following:

- the additively manufacturing orthosis for paraplegic persons is a viable concept – the patient was able to stand and walk, the orthosis was given to him to further train walking and give more feedback,
- automated 3D scanning and scan processing allowed for quicker preparation of design variants, although automation of the design itself would be quite difficult, especially that the patient is an extraordinary case (usually patients with spina bifida are bound to wheelchairs),
- manufacturing of large-scale objects of nylon on a conventional, Cartesian 3D printer takes considerably longer time than of other materials available in the Delta printer – time of manufacturing full nylon orthosis would be approximately of one week, if using only a single machine, while of PET-G it would be possible to deliver a ready orthosis in 2 days,
- two-piece variant failed regardless of material, proving that this concept needs further refining and strength tests,
- additive manufacturing of orthoses for teenagers and adults is possible only using machines of large working chamber – the Delta type machines seem very well suited for the task,

Another, final design iteration is currently in the making – to counteract the leaning of the patient (which is the most problematic issue with the last iteration), the orthosis shape has been slightly rotated. Manufacturing and testing will be a subject of further work.

4 Conclusions

The studies have proven that it is possible to make a 3D printed orthosis for patients with spina bifida to enable them to stand and walk. Quick processing of 3D scans and data extraction for 3D modelling is crucial in preparing subsequent design iterations. For teenagers and young adults the manufacturing process is longer than the usual several hours for the child-sized orthoses [18] and it can take two or three days to complete. However, this is still much less time-consuming and labor-consuming than the traditional manual process and is also much cheaper. Cost of the final orthoses was less than 500 PLN per piece, combining material use, machine working time and human technician post processing time.

Further work will be focused on making the orthosis fully operational, as well as testing it during dynamic activities performed by the patient in the water. More patients will be supplied with the design and it will be fully automated, to allow quick testing for a greater group of patients. Moreover, the methodology of strength testing is currently being devised and the 3D printed orthoses will be subjected to experimental strength testing, to reduce the number of iterations in future work.

Acknowledgements. The studies were realized with a support from Polish National Center for Research and Development, in the scope of the "LIDER" program (grant agreement no. LIDER/14/0078/L-8/16/NCBR/2017). The studies were approved by an appropriate ethical committee.

References

1. Baronio, G., Volonghi, P., Signoroni, A.: Concept and design of a 3D printed support to assist hand scanning for the realization of customized orthosis. Appl. Bionics Biomech. **2017**, 8171520 (2017). https://doi.org/10.1155/2017/8171520
2. Andringa, A., van de Port, I., Meijer, J.W.: Long-term use of a static hand-wrist orthosis in chronic stroke patients: a pilot study. Stroke Res. Treat. **2013**, 546093 (2013). https://doi.org/10.1155/2013/546093
3. Mavroidis, C., et al.: Patient specific ankle-foot orthoses using rapid prototyping. J. NeuroEng. Rehabil. **8**(1), 1–14 (2011). https://doi.org/10.1186/1743-0003-8-1
4. Palousek, D., et al.: Pilot study of the wrist orthosis design process. Rapid Prototyp. J. **20**(1), 27–32 (2014). https://doi.org/10.1108/RPJ-03-2012-0027
5. Paterson, A., Bibb, J.R., Campbell, R.I., Bingham, G.A.: Comparing additive manufacturing technologies for customised wrist splints. Rapid Prototyp. J. **21**(3), 230–243 (2015)
6. Otawa, N., et al.: Custom-made titanium devices as membranes for bone augmentation in implant treatment: modeling accuracy of titanium products constructed with selective laser melting. J. Craniomaxillofac. Surg. **43**(7), 1289–1295 (2015)
7. Jacek, B., et al.: 3D printed models in mandibular reconstruction with bony free flaps. J. Mater. Sci. Mater. Med. **29**(3), 1–6 (2018). https://doi.org/10.1007/s10856-018-6029-5
8. Kaye, R., et al.: Three dimensional printing: a review on the utility within medicine and otolaryngology. Int. J. Pediatr. Otorhinolaryngol. **89**, 145–148 (2016)
9. Jina, Y., Plotta, J., Chena, R., Wensman, J., Shih, A.: A review of the traditional and additive manufacturing of custom orthoses and prostheses. Proc. CIRP **36**, 199–204 (2015)

10. Baronio, G., et al.: A critical analysis of a hand orthosis reverse engineering and 3D printing process. Appl. Bionics Biomech. **2016**, 8347478 (2016). https://doi.org/10.1155/2016/8347478
11. Huotilainen, E., et al.: Inaccuracies in additive manufactured medical skull models caused by the DICOM to STL conversion process. J. Craniomaxillofac. Surg. **42**, 259–265 (2014). https://doi.org/10.1016/j.jcms.2013.10.001
12. Górski, F., Wichniarek, R., Zawadzki, P., Hamrol, A.: Computation of mechanical properties of parts manufactured by fused deposition modeling using finite element method. In: Herrero, Á., Sedano, J., Baruque, B., Quintián, H., Corchado, E. (eds.) Soft Computing Models in Industrial and Environmental Applications. Advances in Intelligent Systems and Computing, vol. 368, pp. 403–413. Springer, Cham (2015). https://doi.org/10.1007/978-3-319-19719-7_35
13. Pilipovi, A., Raos, P., Šercer, M.: Experimental analysis of properties of materials for rapid prototyping. Int. J. Adv. Manuf. Technol. **40**(11–12), 105–115 (2009)
14. Cha, H.Y., et al.: Ankle-foot orthosis made by 3D printing technique and automated design software. Appl. Bionics Biomech. **2017**, 9610468 (2017). https://doi.org/10.1155/2017/9610468
15. Górski, F., et al.: Automated design of customized 3D-printed wrist orthoses on the basis of 3D scanning. In: Okada, H., Atluri, S.N. (eds.) ICCES 2019. MMS, vol. 75, pp. 1133–1143. Springer, Cham (2020). https://doi.org/10.1007/978-3-030-27053-7_97
16. Li, J., Tanaka, H.: Feasibility study applying a parametric model as the design generator for 3D–printed orthosis for fracture immobilization. 3D Print. Med. **4**(1), 1–15 (2018). https://doi.org/10.1186/s41205-017-0024-1
17. Górski, F., Wichniarek, R., Kuczko, W., Żukowska, M., Lulkiewicz, M.: Experimental studies on 3D printing of automatically designed customized wrist-hand orthoses. Materials **13**(18), 4091 (2020)
18. Górski, F., Osiński, F., Żukowska, M., Wierzbicka, N.: Environmental Impact of Additive Manufacturing for Individual Supplies. In: Tonkonogyi, V., et al. (eds.) Advanced Manufacturing Processes II, pp. 384–393. Springer, Cham (2021). https://doi.org/10.1007/978-3-030-68014-5_38
19. Górski, F., Wichniarek, R., Kuczko, W., Żukowska, M.: Study on properties of automatically designed 3D-printed customized prosthetic sockets. Materials **14**(18), 1–26 (2021)

Methodology of the Rapid Manufacturing of an Individualized Anatomical Model of the Tongue with a Tumor for the Preparation of an Organ Reconstruction Operation

Magdalena Żukowska[1](✉), Renata Jezińska[1,2], Filip Górski[1], Wiesław Kuczko[1], Radosław Wichniarek[1], Jacek Banaszewski[2], and Agata Buczkowska-Andruszko[2]

[1] Division of Production Engineering, Poznań University of Technology, Poznań, Poland
magdalena.zukowska@put.poznan.pl

[2] Department of Otolaryngology, Head and Neck Surgery, Poznan University of Medical Sciences, Poznań, Poland

Abstract. Research on the use of individualized anatomical models imitating soft tissues covers more and more areas of medicine. Currently, scientists are focusing on organs such as the heart, kidneys and liver. Tests involving the soft tissues of the oral cavity, for example the tongue, are much less frequent. The article compares various approaches to the production of anatomical models of the tongue with a lesion. The models are designed to assist the physician in preparation for tumor resection surgery and tongue reconstruction. Materials and technologies were proposed that enabled the production of hard models for case visualization, and models imitating soft tissues that enabled simulated surgery. All phantoms were subjected to a medical evaluation.

Keywords: 3D printing · Medicine · Tongue · Simulative operation · Model

1 Introduction

Rapid manufacturing technologies, commonly known as 3D printing, have significantly spread over the last 20 years and are now one of the standard manufacturing methods, alongside methods such as subtractive manufacturing or casting. Therefore, the use of 3D printing is currently visible in most areas of industry or services, as well as medicine. Particularly, several larger application groups can be mentioned: prosthetics, implantology, pre-operative preparation support, education and bioprinting. One of the greatest advantages of 3D printing in medical applications is the ability to reliably recreate complex shapes of anatomical structures, which translates into the individualization of products used in treatment [1]. It is possible to produce prosthetic socket adapted to the individual shape of the patient's stump or orthoses dedicated to individual patients [2, 3], endoprostheses and implants adjusted to the individual shape and structure of the patient's bones and joints [4], anatomical models of the mandible or kidney, improving

F. Gorski et al. (Eds.): MANUFACTURING 2022, LNME, pp. 45–58, 2022.
https://doi.org/10.1007/978-3-030-99769-4_4

the course of the operation and enabling better preparation for operations as well as performing a simulated operation [5–8] and many others. Personalization increases the safety of the procedure and improves the quality of the patient's treatment.

Otolaryngology is one of the areas of medicine dealing with the diagnosis and treatment of disease problems arising within the head and neck. This range also includes ailments related to the tongue or salivary glands. Cancer of the tongue is one of the most common malignancies in the oral cavity. A tongue tumor can grow relatively quickly and spread deeper into the muscles of the tongue. It is recommended that the resulting lesions undergo tumor resection. This procedure involves radical removal of the tumor, and the operation is highly demanding due to the difficulty of accessing the tumor, it may lead to trismus or massive bleeding. After tumor resection, a cavity remains, which should be reconstructed using the patient's tissues. Failure to accurately reconstruct the language is associated with the following problems such as difficulty swallowing, problems with articulation, which can lead to psychological problems [9].

Therefore, individual physical models, representing the distribution of pathology in the area of the tongue are a useful and important element of preoperative preparation. They allow to get acquainted with the location of the lesion, its location and structure as well as the relationship between change and healthy tissue [9]. The correct selection of materials and manufacturing technology is important because it affects the shape and dimensional accuracy, the time obtaining the final model, the costs of obtaining the phantom and its functionality. Models made of materials imitating the hardness of the soft tissues that build the tongue and materials with a noticeably different hardness for the tumor, can be used as an aid in the planning stage of the operation. Also can be used as a model for a simulated operation. During surgery the doctor learns how to properly conduct the incision and how much area must be resected for the operation to be performed correctly.

As part of the research, two approaches to creating tongue models with lesions were proposed. They are divided according to their functions. Models whose only function is visualization were fully produced using the FDM (Fused Deposition Modeling) method. In the case of models that were used to perform the simulated operation, two manufacturing methods were used - FDM and Vacuuum Casting/Molding. The aim of the article is to present the methodology of producing anatomical models of the tongue with a lesion that must undergo resection. Various approaches and the related functionality of the phantom will be compared as well as the medical evaluation of the obtained results.

2 Materials and Methods

2.1 Methodology

Based on medical imaging, it is possible to perform segmentation by the method of e.g. threshold. The files obtained in the study in the DICOM (Digital Imaging an Communications in Medicine) format are processed in a program dedicated to work with medical imaging (including InVesalius, 3D Slicer, Mimics). In the case of tumors of the tongue, two elements are distinguished: the tumor and the tongue. The obtained digital models in the STL format require additional post processing consisting in the removal of artifacts formed during segmentation, smoothing and closing the holes formed in the mesh in order to obtain a waterproof model. All these procedures can be performed in programs that enable work on a mesh (e.g. GOM Inspect, Mesh-mixer, MeshLab) [10].

At this stage, the design and manufacturing procedures are separated depending on the selected strategy. For models whose task is only to visualize the location of the lesion and its relationship with healthy tissue, the files are already imported at this stage into the software dedicated to the selected 3D printer. This procedure is also performed when is access to high-budget devices, which enable simultaneous printing from materials of different hardness and mechanical properties within one object. Uploaded files are sliced and Gcode is generated, which is sent to the printer and the physical model is created. Depending on the printing method selected, the object may require post processing before handing the product over to doctors. Usually it consists removing supports, grinding, trimming or gluing.

The second approach is to combine two manufacturing techniques: additive manufacturing using low-cost equipment (e.g. FDM) and casting using silicone or resin materials. By combining these two technologies, it is possible to obtain physical models made of materials with different mechanical properties and hardnesses. The phantom produced like this is cheaper than using high-cost additive manufacturing methods (e.g. Polyjet). The digital model is used to obtain a negative of the casting mold designed separately for the tumor and for the entire model i.e. tongue+tumor. During designing molds it is important to include the components necessary for proper casting - filler and overflow and the correct dividing line, which will facilitate the subsequent demoulding procedure. It is also important to correctly arrange the negative which help to correctly place the tumor. All procedures related to the design of the mold can be performed in CAD programs and/or programs that enable work on a triangle mesh (e.g. Catia, Blender, Meshmixer). The obtained models are sent to programs dedicated to the printers used, they are sliced and then 3D printed. The manufactured forms require appropriate preparation for the further casting procedure. The casting process takes place as standard moulding, it can also be performed under vacuum conditions (Vacuum Casting) or by using the vacuum chamber only to degass the material. In a situation where the model consists of more than one element, the smaller components are first cast and after they are placed in the final form. Then the final physical model is cast. After the material is hardened, the model requires post processing consisting in the removal of excess material appearing at the edges of the mold connection and around the filler and overflow area. The product prepared in this way can be deliver to doctors. The methodology of designing and producing anatomical models is presented in the diagram below (Fig. 1) [11–15].

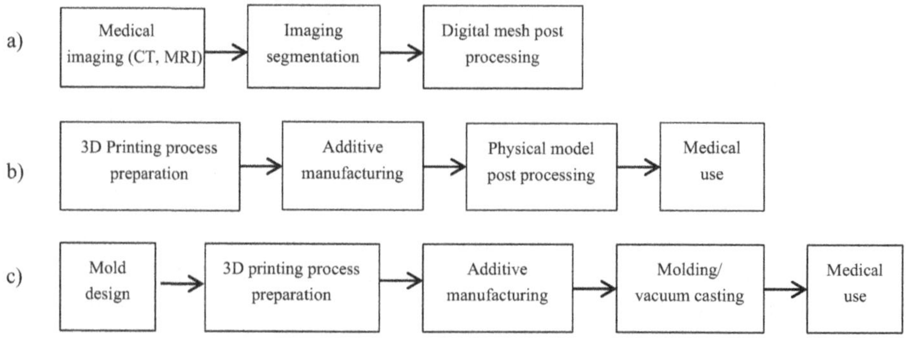

Fig. 1. The methodology of producing anatomical models, a) the procedure of obtaining digital models, b) 3D printing of models directly from the STL file c) the design of casting molds and manufacturing using two different technologies

2.2 Case

The patient was a middle-aged woman with lesion covering the right side of the side of the tongue and partially reaching the median groove of the tongue. As part of the preoperative preparation, medical magnetic resonance imaging (MRI) was performed and the series of pictures was obtained before and after the contrast application. The slice thickness for the images was 1–2 mm.

2.3 Model Designing

The received files were imported into the open source program - InVesalius. In cooperation with a radiologist, manual segmentation of the n lesion was performed using the thresholding method. Then, the segmentation of the tongue was performed in the same way (Fig. 2). However, with a view to further design, including the creation of a negative of the mold, the segmentation of the tongue consisted in marking the correct area for the anatomical shape of the tongue. This means that the mask applied to the image also covered the tumor area at the outline of the tongue. Such a strategy allowed for the creation of elements correctly matched to each other in the next stages. The 2D masks were transformed into 3D digital models, which were exported in STL format.

The obtained models of the tongue and the tumor were smoothed and cleaned of single, unconnected triangles and artifacts. The procedure was performed using the available mesh editing tools in the GOM Inspect 2017 program. The results obtained are shown in the figure (Fig. 3).

The further design stage depended on the function of the model and the selected manufacturing technology. As part of the research the models were made in two different ways.

The first one was to design a fold-out model with easy access to the space in which the tumor was embedded. In this way, the procedure of placing the tumor in the tongue and removing it was simplified in order to get to know the location and position of the tumor in the the tongue. For this purpose, the model was divided into parts. A smaller part was a tongue fold covering the tumor and building it in from the side edge. The larger

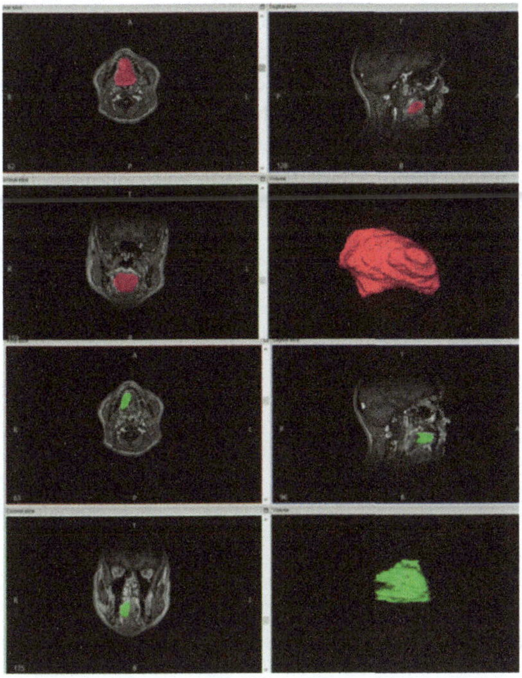

Fig. 2. Preview on windows in InVesalius program: view on masks and 3D models

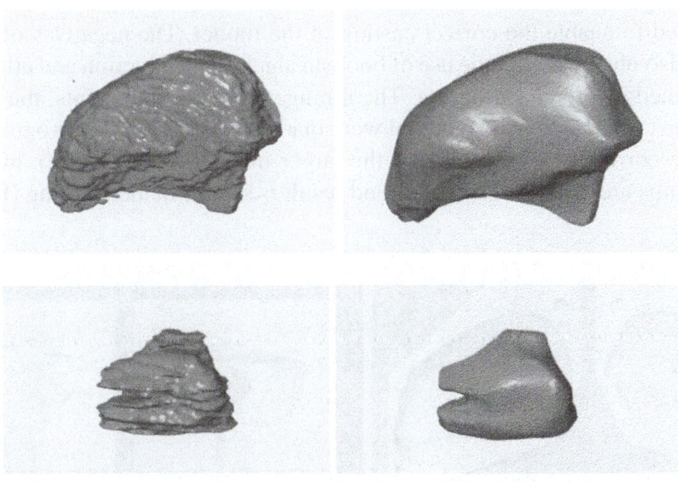

Fig. 3. Models before (left) and after (right) digital post processing

part comprised the rest of the tongue and a lump-like groove. In order to get the correct tumor matching to the tongue, the files were imported into Blender program. As digital models maintain the common origin of the coordinate system during segmentation, the components are positioned in relation to each other in accordance with reality. Therefore, it was possible to perform the subtraction procedure by Boolean Algebra. In this way, a model of the tongue with a tumor-shaped excision was obtained. The further division of the tongue model and the design of the connecting elements were also done in Blender. Finally, the edges of the hole for the tumor were rounded, thus improving the printout and the procedure for placing the element in the model. The obtained effect is shown in the figure (Fig. 4).

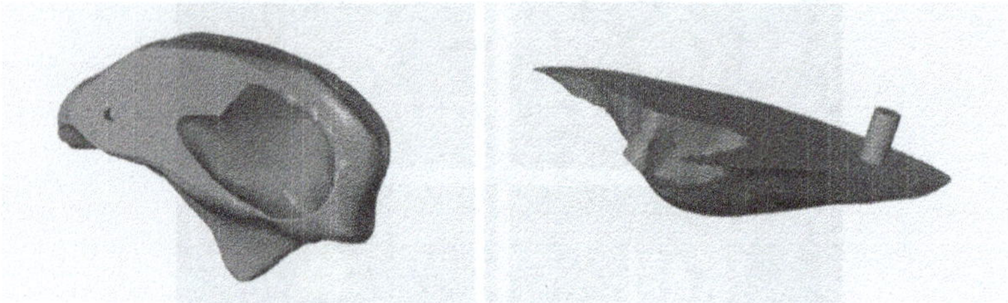

Fig. 4. View of the model split and connecting components

The second approach was to design two molds that were made by additive manufacturing methods. The first form was a tumor negative with additionally arranged holes for filler and overflow. The second form was the negative of combining the tongue and tumor models (reusing the boolean algebra - addition). Also in it, additional components were designed to enable the correct casting of the model. The negatives of forms and holes were also obtained with the use of boolean algebra - subtraction and all procedures were performed in Blender program. The arrangement of components, the division of forms and the cutting of excess material were made in the Meshmixer program. In order to obtain the correct fit of the parts of the larger mold (for the tongue), holes for the connecting pins are also designed. The end result is shown in the drawing (Fig. 5).

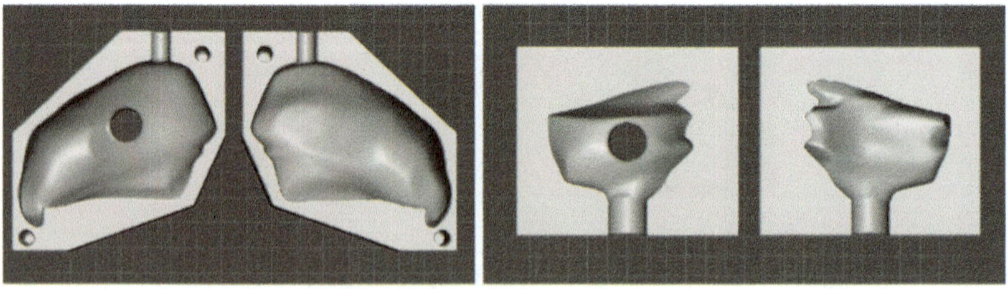

Fig. 5. View of the casting molds for the entire model (left) and the nodule (right)

2.4 Materials and Technologies

Planning the selection of material is closely related to the selection of a specific manufacturing technology. As part of the research, two manufacturing methods were selected: FDM for the production of a demonstrative, fold-out model of the tongue with a tumor and molds and casting using a vacuum chamber to produce semi-transparent models, imitating the hardness of real tissues.

FDM models were made on 3 devices. The fold-out model of the tongue was made of PLA material on the Creality Flash Forge Creator Pro printer, while the tumor was made of ABS material on the Makerbot Replicator 2× printer. All elements were manufactured with a thickness of 0.2 mm and a filling of 25%. The printing time of the tongue model was ~120 min, while the tumor phantom was ~50–60 min. All elements required manufacturing supported by a support material and additionally raft was used to increase the surface of the model's contact with the working platform. In addition, using the FDM method, one of the tumor variants was produced, which was then placed in a casting mold. Flexible TPU material was used to produce this element and it was printed with a Pursa i3 MK3S+printer. The layer thickness was 0.15 mm and the filling was 15%. The printout took ~30 min. All molds were produced in one process. The printout lasted ~540 min, all elements were made of PLA material, the layer thickness was 0.15 mm and the filling was 25%. An FDM Pursa i3 MK3S+printer was also used.

The procedure of casting the tongue and tumor models was divided into two stages. In the first one, tumor models were cast, while in the second one, the form of the tongue was cast with a model of the lesion placed in it. The materials to imitate the tongue and the tumor were selected experimentally based on the prior assessment of the samples by specialist doctors. The material dedicated to the tongue remained the same in all cases and was Sorta Clear 12 (12 Shore A) silicone, while the casting parameters were changed for each model. In the case of the tumor, materials from the given hardness range were selected, which made it possible to compare them with each other and estimate which material best imitates the hardness of the lesion. These materials were: TPU Fiberflex (40 Shore D), silicones: Dragon Skin 10 Very Fast (10 Shore A) and XTX 45 Dry (43 Shore A). Thus, 3 tongue models with the following parameters of of manufacturing were obtained:

- **Model 1:** tongue - Sorta Clear 12 non-degassing silicone + tumor - TPU Fiberflex, molded outside the vacuum chamber
- **Model 2**: tongue - non-degassing and colored Sorta Clear 12 silicone + tumor - Dragon Skin 10 Very Fast, casting outside the chamber and then degassing the final model in vacuum chamber
- **Model 3**: tongue - degassed Sorta Clear 12 silicone + tumor - XTX 45 Dry, molded outside vacuum chamber

Due to the significant coverage of the tumor with the tissues of the tongue, it was not possible to create a cavity in the mold that would facilitate the placement of the tumor model. Therefore, it was necessary to stabilize the position of the model in the negative to prevent the element from moving during casting. In the case of the tumor model made of TPU material, modeling clay was used to place the element inside the mold. Tumor models made of silicone could not be place in this way due to the properties of silicone and its addition to other surfaces. Therefore, the element was set with a use of silicone with shirt cure time (Dragon Skin 10 Very Fast), which allowed for sufficient stabilization of the element during the casting procedure. The material was molded into the mold using a syringe, while in the case of the Dragon Skin 10 Very Fast material, due to its very short pot life time, the casting was made on the open halves of the mold and after pouring them, the molds were joined together, thus joining them into one element.

All the silicones used are two-component and all are mixed in a 1: 1 ratio, with the exception of XTX 45 Dry, which is mixed in a 10:1 ratio. The setting time is different for each material and is presented in the table below (Table 1). The pouring process is ~3–10 min, while the material degassing procedure in the working chamber is ~10 min + ~10 min of rest.

Table 1. List of materials used and their parameters

List of used silicone materials				
Material's name	Hardness [Shore A]	Color	Additional information	
			Pot life [min]	Cure time [h]
Dragon Skin 10 Very Fast	10	Semitransparent	4	0,5
Sorta Clear 12	12	Transparent	40	12
XTX 45 DRY	43	Transparent	90	12

3 Results

3.1 Physical Models

All the models obtained required slight post processing. The entire model of the tongue, produced by the FDM method, had to be cleaned of the supporting material. Due to the printing position, the connecting pin had a very low strength and it was detached quickly. Therefore, the material was melted in this place and a metal pin was inserted in it. Then was stabilized and glued with cyanoacrylate glue. In addition, it was necessary to polish the surface of the tumor to facilitate the placement of the tumor in the tongue model. The finished product was delivered to the doctors (Fig. 6).

The cast models required the removal of excess material accumulating in the places where the mold is joined, as well as the remains of the filler and overflow. In the case of silicone models, some imperfections were noticed, which, depending on the parameters and connections selected, looked different.

In the case of Model 1 (Fig. 7) a significant disturbance of the model transparency was noted due to the filling of the model with air bubbles. In addition, due to the shallow filler, the material supply tube was inserted too deep into mold, which is visible on the final model. Moreover, there was no bonding between the silicone and the TPU material, so it would be possible to freely enucleate the tumor. If the element were not covered with an outer layer of material, the model element could fall out of its own accord.

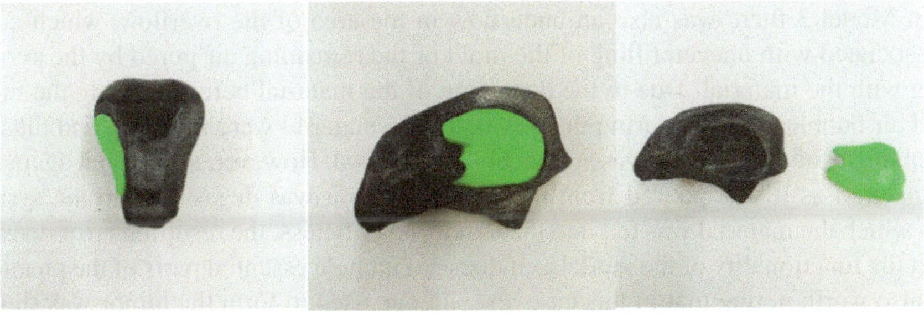

Fig. 6. Fold-out model made of hard plastics

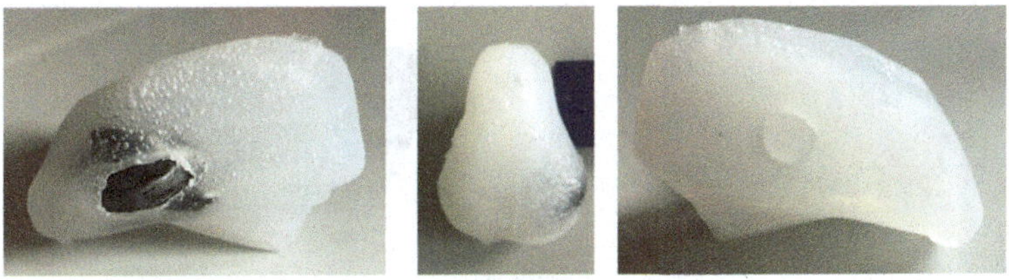

Fig. 7. View of the model number 1

With Model 2 significant underflows across the entire phantom can be noted (Fig. 8). This is related to the degassing method. Namely, the model was degassed after the mold was poured, therefore it was impossible to observe the material sinking and add additional material in the event of a loss of material in the overflow. Additionally, an inflow of air could occur at the moment of rest, already in room conditions, through leaky mold elements, e.g. in the vicinity of the material supply pipe. It is worth noting that the boundaries of the mold connection for the time of degassing were additionally sealed with EVA foam, but not used in the places where the supply pipes were installed. However, the errors that arise do not significantly affect the functionality of the model as they do not cover the resected area. The resulting connection between the silicones was permanent and non-separable. Additionally, in the case of this model, silicone staining was used to build the tongue model to imitate a more realistic situation in which it is not visible through the tissue how deep the change is.

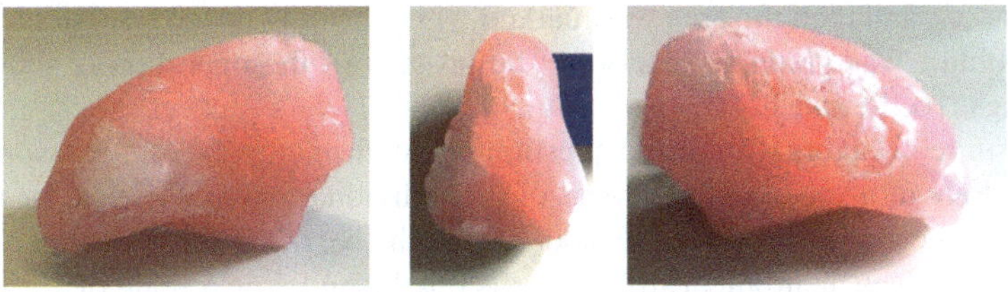

Fig. 8. View of the model number 2

In Model 3 there was also an underflow in the area of the overflow, which could be associated with uneven filling of the mold or the remaining air pured by the syringe along with the material. Due to the degassing of the material before pouring the mold, small air bubbles formed during the mixing of the material were removed and thus the transparency of the material was significantly improved. However, air was put again into the material as it was poured from the vessel where it was degassed into the syringe from which the material was fed into the mold. Nevertheless, the resulting error does not affect the functionality of the model as it does not include essential parts of the phantom. It is also worth noting that in this case, the silicone used to form the tumor was stained to improve the visibility and position of the tumor in the tongue (Fig. 9).

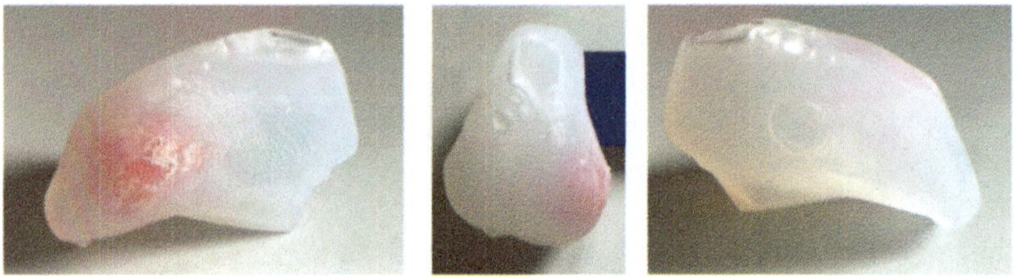

Fig. 9. View of the model number 3

3.2 Medical Assessment

All models have been subjected to medical evaluation by specialist doctors - laryngologists working in the Department of Otolaryngology, Head and Neck Surgery at Poznan University of Medical Sciences.

The visual model, made entirely of high-hardness materials, fully fulfilled its function. The size and shape of the cutout lesion coincided with the produced phantom. The model can assist in the planning of surgery and the approach to tumor removal. Moreover, it allows to estimate the size of tissue collected for tongue reconstruction.

Models for visual assessment and simulated surgery were assessed both in the interview and in the questionnaire, collecting basic information about subjective feelings of doctors. At the beginning, specialists assessed the visual aspects of the models, and then proceeded to perform a simulated operation. First, the tumor margins were determined using a marker, and then the area was marked with a healthy tissue reserve of ~ 10 mm. Then the tumor was gradually excised. The process of the operation is shown in the photos (Figs. 10, 11 and 12).

In the survey, doctors were asked questions about the general feeling of working with models and about a specific assessment of the individual materials used for the study. Two out of three doctors considered Model 3 was the best, while one person considered it was Model 1. The anatomical structure was rated 4 and 5 on a scale of 1 (very bad) - 5 (very good). General feelings about the material used to create the tongue model were assessed by all on 4. Doctors, when asked about general and aesthetic feelings about individual models, assessed them as follows: Model 1 on 3 and 4, Model 2 also on 3

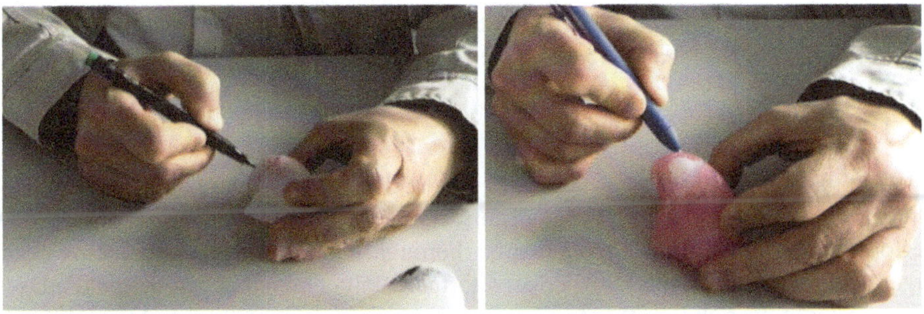

Fig. 10. Marking the boundaries of the tumor and the cutting margin

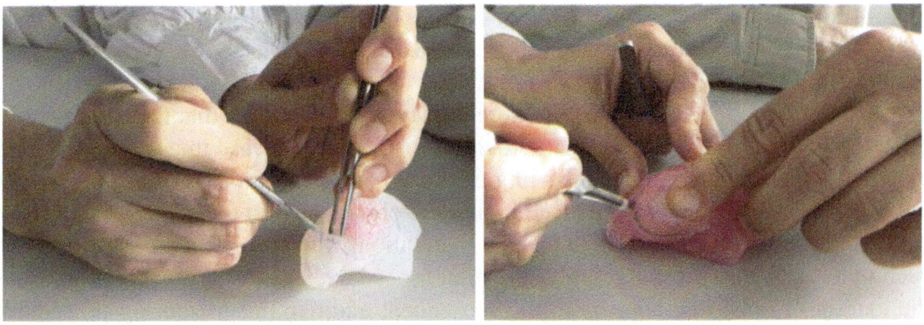

Fig. 11. Tumor resection procedure - simulated operation

Fig. 12. The final result of simulated operation

and 4, while Model 3 on 4 and 5 The least useful in pre-operative preparation would be Model 1, while Model 2 and 3 could be used in the planning of surgery.

4 Discussion

The soft tissue segmentation procedure is a more difficult procedure as it often requires either complete manual work or manual corrections on each layer. In addition, the person performing the segmentation must first be thoroughly familiar with the anatomy of the organ and complement the knowledge with the data contained in radiological atlases. The support of the radiologist in this process increases the quality of the obtained model, however, it extends the time obtaining the ready digital model (the necessity to schedule an appointment with the doctor in the spare hours from work). The quality of the obtained digital models is influenced by the series of photos received in the radiological examination. For soft tissues, the recommended layer thickness is 0.5–1 mm [16]. In this way, it will be possible to create a shape-accurate model and reduce the likelihood of missing significant changes in the tumor-healthy tissue relationship. In addition, imaging containing images after application of contrast can improve the reading of the lesion outline.

Designing models that consist of more than two parts requires that they be properly arranged in relation to each other in order to be able to freely use the phantom. When planning the arrangement of connecting elements, the orientation in which the model will be printed should also be planned in order to obtain the best mechanical properties, especially for small and loaded elements. In addition, when designing models fitted to each other in the boolean algebra subtraction procedure, one should take into account the margin of error that appears in the subsequent additive manufacturing procedure. Designing casting molds requires taking into account the location of the area of interest (around the lesion and tumor) and the distribution of the infusion and overflow in relation to them, so that in the event of a underflow it does not cover the area necessary for planning the operation. In addition, the demoulding procedure should be taken into account. The form must be constructed in such a way that the procedure runs smoothly and does not damage the model. In the tests, the best visual effect was obtained when the material was degassed before the final casting was made. The problem of the resulting deficiency can be eliminated by degassing the material already in the syringe, through which the silicone is fed to the mold. Due to the relatively small volume of the model, it is possible to perform such a procedure.

During the discussion after the simulated operation, the following conclusions can be made. The use of transparent material makes it easier to get to know the size, shape and location of the tumor, which is associated with a better estimation of the margin of tissue surrounding the diseased tissues. The doctor preparing for the operation can also perform a trial procedure on such a model, thanks to which he will estimate the size of the removed tissues and, at the same time, the size of the tissues needed for extraction for further tongue reconstruction. Getting to know the pathology before the operation increases the patient's safety and increases the confidence of the doctor and shorter operation time. The models also have significant educational value, they can be a learning tool for students, trainees and residents who have not had the opportunity to familiarize themselves with specific pathologies before and how to act in certain cases.

In imitation of tissue hardness, silicone materials work best for the tissues of the tongue in the range of 10–15 Shore A and for the tumor 35–45 Shore A. It is important that the difference in hardness is noticeable, because during the actual operation, assessment by touch and based on medical imaging are about scratching the cutting area. Staining the tumor element improves its visibility in the final phantom. On the other hand, the staining of the tongue element increases the level of difficulty during the simulated operation and more reflects the actual state of affairs in the operating room. Such a model can also be used in the education process, when the student is to practice the learned procedure on a transparent model in more realistic conditions.

5 Conclusions

Additive manufacturing methods can be important in medical applications. The areas in which solutions using 3D printing can be implemented are gradually expanding. More and more often they cover the subject of anatomical structures made of soft tissues and attempts are made to imitate these tissues. Depending on the needs of the doctor, patient, available technological solutions, time and budget, individual phantoms can be implemented at the stage of diagnosis and operational planning. Their task is primarily to visualize the problem, but it is possible to create models that improve preparation for surgery by performing an operation simulated on an artificial object. In otolaryngology, the use of such phantoms is still being recognized, and there are few articles researching this topic in scientific databases. Therefore, it is important to check and further study how anatomical models of the tongue with a tumor influence the preparation of doctors and the planning of tumor resection and tongue reconstruction surgery. Further research should investigate clinical implementations, compare the pathology models with the actual sample, and in the longer term, it is worth considering the creation of additional models simulating tissue flaps collected for reconstruction, thanks to which it will be possible to perform a simulated operation not only of tumor resection, but also of tongue reconstruction.

References

1. Żukowska, M., Górski, F., Wichniarek, R., Kuczko, W.: Methodology of low cost rapid manufacturing of anatomical models with material imitation of soft tissues. Adv. Sci. Technol. Res. J. **13**(4), 120–128 (2019)
2. Górski, F., Wichniarek, R., Kuczko, W., Żukowska, M., Lulkiewicz, M., Zawadzki, P.: Experimental studies on 3D printing of automatically designed customized wrist-hand orthoses. Materials **13**, 4091 (2020)
3. Górski, F., Wichniarek, R., Kuczko, W., Żukowska, M.: Study on properties of automatically designed 3D-printed customized prosthetic sockets. Materials **14**, 5240 (2021)
4. Malyala, S.K., Kumar, R., Alwala, A.M.: A 3D-printed osseointegrated combined jaw and dental implant prosthesis – a case study. Rapid Prototyp. J. **23**(6), 1164–1169 (2017)
5. Ganry, L., Quilichini, J., Bandini, C.M., Leyder, P., Hersant, B., Meningaud, J.P.: Three-dimensional surgical modelling with an open-source software protocol: study of precision and reproducibility in mandibular reconstruction with the fibula free flap. J Oral. Maxillofac. Surg. **46**(8), P946-957 (2017)

6. Adams, F., et al.: Soft 3D-printed phantom of the human kidney with collecting system. Ann. Biomed. Eng. **45**, 963–972 (2017)
7. Bernhard, J.-C., et al.: Personalized 3D printed model of kidney and tumor anatomy: a useful tool for patient education. World J. Urol. **34**(3), 337–345 (2015). https://doi.org/10.1007/s00345-015-1632-2
8. Lee, H., Nguyen, N.H., Hwang, S.I., Lee, H.J., Hong, S.K., Byun, S.S.: Personalized 3D kidney model produced by rapid prototyping method and its usefulness in clinical applications. Int. Braz. J. Urol **44**(5), 952–957 (2018)
9. Irace, A.L., et al.: Additive manufacturing of resected oral and oropharyngeal tissue: a pilot study. Int. J. Environ. Res. Public Health **18**, 911 (2021)
10. Gibson, I., et al.: The use of rapid prototyping to assist medical applications. Rapid Prototyp. J. **12**(1), 53–58 (2006)
11. Muguruza Blanco, A., Krauel, L., Fenollosa Artés, F.: Development of a patients-specific 3D-printed preoperative planning and training tool, with functionalized internal surfaces, for complex oncologic cases. Rapid Prototyp. J. **25**(2), 363–377 (2019)
12. Witowski, J.S., Pędziwiatr, M., Major, P., Budzyński, A.: Cost-effective, personalized, 3D-printed liver model for preoperative planning before laparoscopic liver hemihepatectomy for colorectal cancer metastases. Int. J. Comput. Assist. Radiol. Surg. **12**(12), 2047–2054 (2017). https://doi.org/10.1007/s11548-017-1527-3
13. Adams, F., et al.: Soft 3D-printed phantom of the human kidney with collecting system. Ann. Biomed. Eng. **45**(4), 963–972 (2016). https://doi.org/10.1007/s10439-016-1757-5
14. Malyala, S.K., Kumar, R., Alwala, A.M.: A 3D-printed osseointegrated combined jaw and dental implant prosthesis – a case study. Rapid Prototyp. J. **23**(6), 1164–1169 (2017)
15. Górski, F., Wichniarek, R., Kuczko, W., Banaszewski, J., Pabiszczak, M.: Application of low-cost 3D printing for production of CT-based individual surgery supplies. In: Lhotska, L., Sukupova, L., Lacković, I., Ibbott, G. (eds.) World Congress on Medical Physics and Biomedical Engineering 2018. IFMBE Proceedings, vol. 68/1, pp. 249–253. Springer, Singapore (2019). https://doi.org/10.1007/978-981-10-9035-6-45
16. Bibb, R., Winder, J.: A review of the issues surrounding three-dimensional computed tomography for medical modelling using rapid prototyping techniques. Radiography **16**(1), 78–83 (2010)

Thin-Film Protective Coatings on Samples Manufactured by Direct Metal Laser Sintering Technology Used in Dentistry

Diana-Irinel Băilă[1(✉)], Răzvan Păcurar[2], and Ancuţa Păcurar[2]

[1] University Politehnica of Bucharest, Blv. Splaiul Independenţei, no. 313, sector 6, 060042 Bucharest, Romania
baila_d@yahoo.com

[2] Faculty of Machine Building, Department of Manufacturing Engineering, Technical University of Cluj-Napoca, Blv. Muncii, no. 103-105, 400641 Cluj-Napoca, Romania

Abstract. In the last decade, additive manufacturing technologies and especially direct metal laser sintering (DMLS) had become a great sustainability development method which can be widely used in the industry for testing custom-designed materials to create highly complex geometry parts that cannot be made by conventional methods. Dental restorations are frequently made by DMLS due to the accuracy of this technology, which allows the obtaining of customized dental restoration in a short period of time to be implanted to the patient in a single day with low cost, by greatly improving the patient's comfort. The research presents DMLS principle, cycle life of this technology, as well as the coatings with Ni and with hydroxyapatite obtained by sol-gel method.

In this paper there were realized experimental researches concerning the microstructure of hydroxyapatite coating on Co-Cr alloy samples materialized by DMLS process, which were post-treated at the end and Ni coating on Cu-Zn cast alloy. The results obtained were compared with similar data related to the same area of research as these were reported in the literature.

Keywords: Direct metal laser sintering · Sol-gel method · Co-Cr alloy · Cu-Zn alloy · Hydroxyapatite · Ni coating

1 Introduction

1.1 Generalities

The last industrial revolution (Industry 4.0.) implies the integration of smart manufacturing systems and developed information technology. Direct metal laser sintering (DMLS) technology represents an innovative technology that can be used for realizing functional and durable products in different domains, such as airspace, aeronautics, electronics, automotive and medicine, which explains the great interest of the enterprises to implement sustainability and environmental protection for this technology, with lower costs. The development of DMLS technology nowadays makes possible to consider the

F. Gorski et al. (Eds.): MANUFACTURING 2022, LNME, pp. 59–68, 2022.
https://doi.org/10.1007/978-3-030-99769-4_5

manufacturing of metallic structural parts using advanced materials with efficient waste minimization, high design freedom, and highly complex geometries, that are hardly possible to be made using conventional manufacturing methods, such as: undercuts; channels that are passing through sections; tubes which are interconnected within tubes; internal voids [1–3].

Due to this technology, which allows the parts to be manufactured based on layer by layer method, designing parts cannot be limited but should consider the parts' life cycle and the risks of stresses that occur in the material structure that is melted and solidified using the laser (welding process).

Using additive manufacturing requires a combination of skills and means in the fields of metallurgy, mechanics and manufacturing methods. Figure 1 presents additive manufacturing technologies performance concerning surface quality, part size, geometric complexity, manufacturing time, machine availability (number of machines available in the world), static properties, fatigue strength, fabrication cycle in industry, and, as can be noted, the DMLS process assures the best surface quality, geometric complexity, mechanical properties, and machine availability, but cannot be used to obtain large size parts and the manufacturing cycle is low.

In comparison with other additive manufacturing technologies, such as EBM (electron beam melting), LMD (laser projection/laser metal deposition), DMD (direct metal deposition), only SLM (selective laser melting) and DMLS technologies can be used in dentistry.

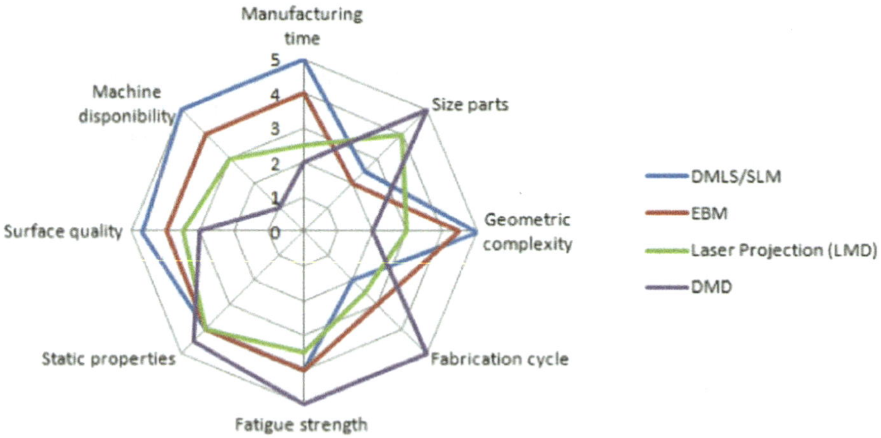

Fig. 1. Additive manufacturing technologies performance.

The DLMS process consists of sintering a powder material using a laser that binds powder grains to materialize a solid structure by welding. The un-sintered or loose material is removed after the sintering process is finished on the machine and can be recycled for future use, making it both economical and environmentally friendly [4–6].

1.2 Life Cycle Analysis of DMLS Technology

Life cycle analysis (LCA) is a method that is related to the environmental aspects with minimal or no adverse environmental impact and the potential impacts throughout the life of a product from the acquisition of the raw material to its production, use, and disposal [7]. The LCA for a Co-Cr alloy dental crown made using DMLS is presented in Fig. 2.The importance of environmental protection and the impact of realized products on the environment in close correlation with the manufacturing technologies and used materials has increased the interest in developing methods to better comprise the impact of all these issues at world level.

One of the techniques under development is LCA. LCA can be used for:

- Identifying the opportunities for improving the environmental performance of products at different stages of their life cycle; [8].
- Establishing decision procedures in industry, government, and non-governmental organizations,
- Choosing relevant environmental performance indicators;
- Marketing.

The DMLS process implies long-term competitiveness by optimizing labor, energy, and materials to produce a high-quality products and to find a rapid response for variation in market demands and delivery time of the products.

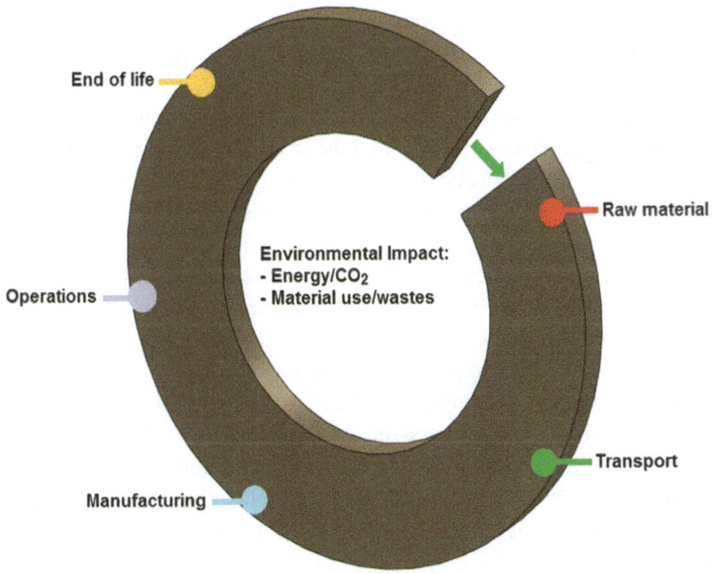

Fig. 2. Life cycle analysis of a dental crown made by Direct Metal Laser Sintering.

DMLS is a sustainable manufacturing method because it allows the realization of dental restoration for customized products, starting from patient radiography or CT (Computer tomography), with high precision and very good quality.

After the parts are finished on the machine, they do not require any adjustment or post-processing procedures; they can be simply directly mounted in the patient's mouth.

When the DMLS method is used, parts can be produced with low manufacturing time and a dental operation can be completed in a single day. Using this technology, material waste is very small because dental restorations do not depend on any cutting tool, and manufacturing repeatability is very good [9–11].

2 Results and Discussion

2.1 Materials and Methods

The most used technologies for the manufacturing metallic powders for DMLS are the hydride-dehydride process, water atomization, gas atomization, plasma atomization, electrode induction melting gas atomization (EIGA), plasma rotating electrode process (PREP), centrifugal atomization and plasma spheroidization [10].

The most common method of metal powder production for DMLS is plasma atomization. The elemental feedstock is melted under an air or inert gas blanket or under vacuum. The chamber is then backfilled with gas to force molten alloy to pass through a nozzle where high-velocity air, N, He, or Ar gas impinges the powder grains onto the flowing melt and breaks it up.

The powder is mostly spherical, with some asymmetric particles and satellites present. A satellite is when a smaller particle sticks to a larger one during solidification. Grain size ranges from 0 to 500 microns.

The most used powders that can be produced using this method are based on Ni, Co, Fe, Ti, and Al alloys [11–17].

Titanium cannot be directly brought in the powder state because it presents increased ductility compared to other crystallized metals, crystallinity being defined by the ratio of crystalline c/a over the theoretical value 1.633, which allows splitting grains after multiple plans with an atomic density close to the base or median plane [12, 17].

Titanium is highly reactive to warm and oxide refractory materials so that the production of powders by mechanical means has the advantage of allowing the possibility to keep it at high purity level without being necessary to take special measures and expensive protection measures.

The technological method required for reaching the titanium powder by hydride-milling-de-hydride systems is strongly correlated to the reaction of titanium to hydrogen.

The solvus curve in the diagram is steep and provides the diffusion of a large amount of hydrogen in the solid solution [12, 17]. Upon heating under equilibrium conditions, for temperatures above 450 °C, hydrogen forms a metastable hydride—TiH_2.

This can be brought to ambient temperature in the atmosphere without decomposition or by a complex reaction.

It is extremely brittle and may be easily sieved to give a particle size fraction distribution range. In the case of vacuum heating, the hydride decomposes, liberating hydrogen [18–20].

Characteristics of the powder particles at the end of the manufacturing process, using different standards, are shown in Table 1.

Table 1. Standards (ASTM, ISO, EN) for powder properties used in additive manufacturing [14–16]

AM Powder characteristics	Powder type	Symbols	Techniques	ASMT standard	ISO standard	EN standard
Size and shape	Metallic powders	Φ [μm]	SEM	B822	13322	–
Specific density	Metallic powders	$\rho_{specific}$ [g/cm^3]	Gas pycnometer	B293	12154	–
Apparent density	Non-free flowing metallic powders	ρ_{app} [g/cm^3]	Hall apparatus	B212	3923/1	3923
Apparent density	Non-free flowing metallic powders	ρ_{app} [g/cm^3]	Carney apparatus	B417	3923/1, 4490	4490
Apparent density	Metallic powders	ρ_{app} [g/cm^3]	Arnold meter	B703	–	–
Apparent density	Refractory metals and compounds	ρ_{app} [g/cm^3]	Scott volumeter	B329	3923/2	–
Tap density	Metallic powders	P_{tapped} [g/cm^3]	BT-1000	B527	3953	3953
Average particle size	Metallic powders	d_{60}	Fisher sub-sieve sizer	B330, C72	10070	–
Powder sieve analysis	Metallic powders	–	Sieve analysis equipment Westmoreland	B214	4497,2591	24497
Particle size distribution	Metallic powders and related compounds	d_{10}, d_{60}, d_{90}	Light scattering	B822	13320, 24370	–
Flowing rate	Free-flowing metallic powders	Flow time (s) for 50 g	Hall apparatus	B213	4490	4490
Envelope specific surface	Powder bed under steady flow	S_v [m^2/g]	Measurement of air permeability	–	10070	196-6

Table 2. Mechanical characteristics of Co-Cr powder [12, 28]

Minimum layer thickness	20 μm
Surface roughness	Ra=10 μm, Ry=40-50 μm Ra=0,39 μm, Rz=1,6 μm After polishing Rz<1 μm
Density with standard parameters	8,3 g/cm^3
Mechanical properties	
Tensile strength	1100MPa
Yield strength	600 MPa
Elongation at break	20%
Young's modulus	200 GPa
Hardness	35-35 HRC
Fatigue life	>10 million cycles
Thermal properties	
Maximum operating temperature	1150 ^{0}C

The SINT-TECH company and ISO 9001 and ISO 13485 standards propose a range of powders suitable for the process developed by Phenix Systems.

This powder range was selected to guarantee an optimized result for implementation with the "PX" range and the former "PM" range systems produced by Phenix Systems [18–22].

The Co-Cr alloy powder (ST2724G) which is suitable for the DMLS manufacturing comprises the following chemical composition: 54.31% Co; 23.08% Cr; 11.12% Mo, 7.85% W, 3.35% Si, and Mn, Fe < 0.1% [23, 24]. Table 2 presents the mechanical characteristics of the Co-Cr powder used for the DMLS manufacturing process.

The DLMS process consists of sintering a powder material using a laser that binds powder grains to materialize a solid structure by welding. The un-sintered or loose material is removed after the sintering process is finished on the machine and can be recycled for manufacturing of other parts by DMLS in the future [4, 6].

The Co-Cr alloy powder has low granulometry, and the size of the spherical grain (average dimension) is approximately 20 μm [25].

For the Cu-Zn cast alloy samples, the chemical composition was 68.09 wt% Cu and 31.91 wt% Zn.

2.2 Direct Metal Laser Sintering Process

DMLS process is used in different domains, such as aerospace (fuel injection, structural elements, blades), tooling (tooling inserts), automotive (air ducts, Formula 1 components), medical domain (crowns, analog implants, copings, artificial hip joints, medical instruments) [26, 27].

In the series production manufacturing readiness level by Rolland Berger, the DMLS process in the dental domain has readiness level 10; this is full-rate production compared to other industries [26, 27].

The quality of the final product made by DMLS depends on the following key parameters:

- Material powder; this is the responsibility of the ALM manufacturer in terms of chemical composition, grain size, and grain shape;
- Process parameters, such as laser power, scanning speed, layer thickness, atmosphere chamber, and temperature in the DMLS process, etc.
- Part geometry, orientation, build orientation in close connection to the shape of the part (thickness variation, massive area) and position and number of parts in the chamber, etc.

For experimental research, the samples were designed by SolidWorks and were saved as "*.stl" files. The samples were sintered using a Phenix Systems machine type PXS & PXM Dental. The machine uses a fiber laser with the following characteristics: laser power $P = 50$ W, wavelength $\lambda = 1070$ nm.

The total working space of the machine is $100 \times 100 \times 80$ mm. The temperature used for the sintering process was 1300 °C. Nitrogen gas was used inside the working chamber of the machine for this process [27].

To obtain very good quality dental crown surfaces nitrogen gas and argon gas were used, especially in the case when industrial parts are produced using this type of machine.

The personalized conventional dental crown of Co-Cr are manufactured by DMLS, after patient radiography, which are used for shortening the dental surgery time, leading to faster healing and recovering of daily life for the patient, representing a simple and eloquent sustainability process [28].

2.3 Thin-Film Protective Coatings

Before performing of coating deposition, all DMLS samples were required to be washed with distilled water and to be treated with alcohol. In this research, the procedure for performing the sol-gel synthesis was similar to the one that has been reported by Fathi and Hafini [29]. The phosphoric pentoxide (P2O5, Merck) was dissolved in absolute ethanol to form a 0.5 mol/L solution and calcium nitrate tetrahydrate [Ca(NO3)2.4H2O,Merck] was also dissolved in absolute ethanol to obtain a 1.67 mol/L solution. The solutions were mixed in a molar ratio of $Ca/P = 1.67$ and were homogenized by using a magnetic stirrer for 48 h, to materialize the sol. By using the dip coating method, the hydroxyapatite has been deposited on Co-Cr alloy disks samples manufactured by DMLS process.

After coating deposition has been performed, the samples were dried for a few minutes and they were subjected to a thermal treatment which was performed using an electrical furnace, in air atmosphere. At the end, the samples were left to cool slowly to the ambient temperature.

In Fig. 3 is presented sol-gel hydroxyapatite on Co-Cr alloy sample, sintered by DMLS process and in Fig. 4 is presented Ni coating on Cu-Zn cast alloy sample. In the Fig. 4, the coating was realized by physical vapor deposition.

Both coatings present a good uniformity and a homogeneous chemical composition, the substrates being realized of Co-Cr alloy, respectively Cu-Zn alloy having good mechanical properties, and the coatings presented a good adhesion to the substrate, improving bioactivity and corrosion resistance. Both types of materials can be used successfully in the dental field.

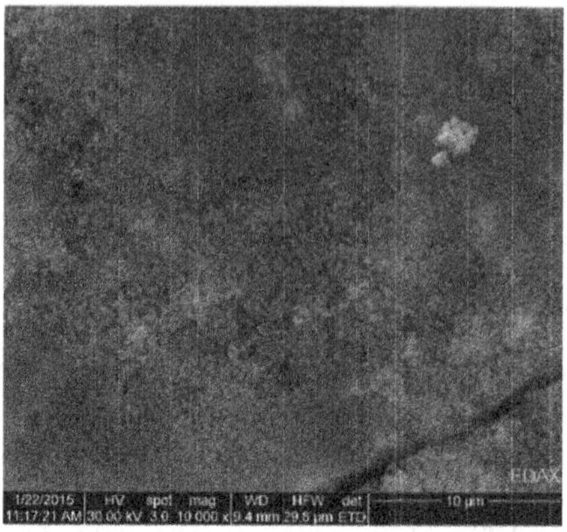

Fig. 3. Sol-gel hydroxyapatite coating on Co-Cr alloy sample sintered by DMLS process (×10000)

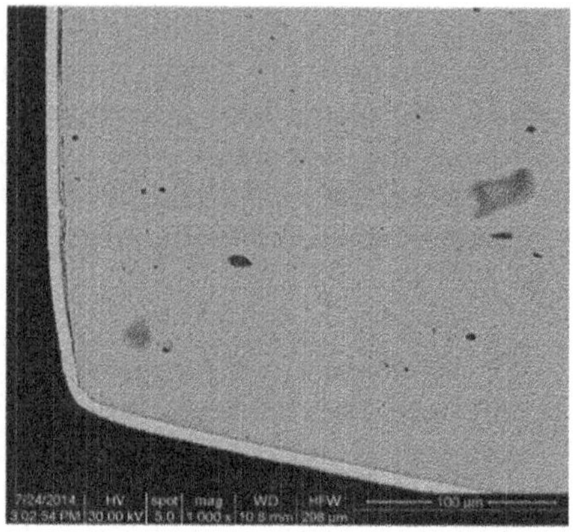

Fig. 4. Ni coating on Cu-Zn cast alloy sample (×1000)

3 Conclusions

The DMLS process has changed the domains of dental restoration manufacturing in many aspects. This type of manufacturing technology presents high flexibility in production for realizing high-quality, personalized medical parts, that are identical with the patient radiography or CT. The DMLS technique has produced a real industrial revolution in the dental domain. The sustainability of DMLS is distinguished by the fact that dentists and dental technicians work much easier thanks to this technology; the patient is also treated in a much shorter time due to the possibility of quickly manufacturing customized dental elements.

The appearance and use of DMLS technology in medicine has led to many innovations and opportunities in the medical domain and allows the decreasing costs in close connection with time-savings for surgical operations in the dental domain by eliminating the post-processing operations that are necessary in the case of dental restorations produced using conventional technologies. DMLS has several advantages but also presents some limits, such as materials limitations, chamber build limitations, laser power limitations, the inert atmosphere required for the manufacturing process, machine cost, and production cost.

DMLS process is a reliable technology that can be used in dental and medical applications, having a full rate of production, but for other domains, such as tooling, aerospace, and automotive, more sustainable development research is required to increase the accuracy and efficiency of DMLS with respects to the decreasing material and manufacturing system costs, which will allow much faster production of products.

The research in this paper concerning the chemical coating with hydroxyapatite on Co-Cr alloy sintered by DMLS and the physical vapor deposition with Ni on Cu-Zn cast alloy. Both samples present the uniformity of micro-structures, using SEM analysis, but porous structure for the HA coating and a compact structure for Ni coating. HA coating was realized to improve the bioactivity for the dental implants, and Ni coating was realized to improve the corrosion and wear resistance.

Coatings obtained by chemical methods are more durable and can be made on larger parts surfaces, compared to physical coatings, which are more expensive and allow the coating of small surfaces with very fine and uniform layers, of the order of nanometers.

In the future, additive manufacturing will evolve in developing of new types of innovative technologies, such as 4D printing, 3D bioprinting, and hybrid manufacturing, thanks to the advances that are performed in this field from one year to another and thanks to the intelligent materials that are developed year by year, allowing the possibility of developing more complex and multifunctional components not only in industry but also in the medical field.

References

1. Khairallah, S.A., Anderson, A.T., Rubenchik, A., King, W.E.: Laser powder-bed fusion additive manufacturing: physics of complex melt flow and formation mechanisms of pores, spatter, and denudation zones. Acta. Mater. **108**, 36–45 (2016)
2. Mehrpouya, M., et al.: The potential of additive manufacturing in the smart factory industrial 4.0: a review. Appl. Sci. **9**, 3865 (2019)
3. Matthews, M.J., Guss, G., Khairallah, S.A., Rubenchik, A.M., Depond, P.J., King, W.E.: Denudation of metal powder layers in laser powder bed fusion processes. Acta Mater. **114C**, 33–42 (2016)
4. DebRoy, T., et al.: Additive manufacturing of metallic components process, structure and properties. Prog. Mater. Sci. **92**, 112–224 (2018)
5. Ford, S., Despeisse, M.: Additive manufacturing and sustainability: an exploratory study of the advantages and challenges. J. Clean. Prod. **137**, 1573–1587 (2016)
6. Lapcevic, A.R., Jevremovic, D.P., Puskar, T.M., Williams, R.J., Eggbeer, D.: Comparative analysis of structure and hardness of cast and direct metal laser sintering produced Co-Cr alloys used for dental devices. Rapid. Prototyp. J. **22**, 144–151 (2016)
7. Zheng, L., McMahon, C., Li, L., Ding, L., Jamshidi, J.: Key characteristics management in product lifecycle management: a survey of methodologies and practices. Proc. Inst. Mech. Eng., Part B: J. Eng. Manuf. **222**(8), 989–1008 (2008)
8. Kellens, K., Renaldi, R., Dewulf, W., Kruth, J.P., Duflou, J.R.: Environmental impact modeling of selective laser sintering processes. Rapid Prototyp. J. **20**, 459–470 (2014)
9. American Center for Life Cycle Assessment: LCA XIV International Conference, pp. 130–141. CA, United States, San Francisco (2014)
10. Kellens, K., Mertens, R., Paraskevas, D., Dewulf, W., Duflou, J.: Environmental impact of additive manufacturing processes: does am contribute to a more sustainable way of part manufacturing. Proc. CIRP **61**, 582–587 (2017)
11. Dawes, J., Bowerman, R., Trepleton, R.: Introduction to the additive manufacturing powder metallurgy supply chain. Johnson Matthey Technol. Rev. **59**(3), 243–256 (2015)
12. Slotwinski, J.A., Garboczi, E.J., Stutzman, P.E., Ferraris, C.F., Watson, S.S., Peltz, M.A.: Characterization of metal powders used for additive building. J. Res. Natl. Inst. Stand. Technol. **119**, 460–494 (2014)
13. https://www.carpenteradditive.com/technical-library/. Accessed 09 Jan 2020
14. Standards ASTM B822, ASTM B293, ASTM B212, ASTM B417, ASTM B703, ASTM B329, ASTM B527, ASTM B330, C72, ASTM B214, ASTM B822, ASTM B213
15. Standards ISO13322, ISO12154, ISO3923/1, ISO4490, ISO3923/2, ISO3953, ISO10070, ISO4497, ISO2591, ISO13320, ISO24370, ISO13485, ISO9001

16. Standards EN3923, EN4490, EN3953, EN24497, EN196-6
17. Tang, H.P., Qian, M., Liu, N., Zhang, X.Z., Yang, G.Y., Wang, J.: Effect of powder reuse times on additive building of Ti-6Al-4 V by selective electron beam melting. JOM **67**(3), 555–564 (2015)
18. Thijs, L., Verhaeghe, F., Craeghs, T., Van Humbeeck, J., Kruth, J.P.: A study of the micro structural evolution during selective laser melting of Ti-6Al-4V. Acta Mater. **58**, 3303–3312 (2010)
19. Băilă, D.I., Doicin, C.V., Cotruţ, C.M., Ulmeanu, M.E., Ghionea, I.G., Tarbă, C.I.: Sintering the beaks of the elevator manufactured by direct metal laser sintering (DMLS) process from Co-Cr alloy. Metalurgija. **55**, 663–666 (2016)
20. Drstvensek, I., et al.: Applications of rapid prototyping in Cranio-maxilofacial surgery procedures. Int. J. Biol. Biomed. Eng. **1**, 29–38 (2018)
21. Atzeni, E., Iuliano, L., Minetola, P., Salmi, A.: Proposal of an innovative benchmark for accuracy evaluation of dental crown manufacturing. Comput. Biol. Med. **42**, 548–555 (2012)
22. Pasebani, S., Ghayoor, M., Badwe, S., Irrinki, H., Atre, S.V.: Effects of atomizing media and post processing on mechanical properties of 17-4PH stainless steel manufactured via selective laser melting. Addit. Manuf. **22**, 127–137 (2018)
23. Béreš, M., et al.: Mechanical and phase transformation behaviour of biomedical Co-Cr-Mo alloy fabricated by direct metal laser sintering. Mater. Sci. Eng. A. **714**, 36–42 (2018)
24. Barucca, G., et al.: Structural characterization of biomedical Co-Cr-Mo components produced by direct metal laser sintering. Mater. Sci. Eng. C. **48**, 263–269 (2015)
25. Girardin, E., et al.: Biomedical Co-Cr-Mo components produced by direct metal laser sintering. Mater. Today. **3**, 889–897 (2016)
26. Jacob, G., Brown, C., Donmez, A., Watson, S., Slotwinski, J.: Effects of powder recycling on stainless steel powder and built material properties in metal powder bed fusion processes. In: NIST Advanced Building Series, pp. 100–106 (2016)
27. Building Readiness Level for ALM (Additive Laser Building) by Rolland Berger, https://docplayer.net/2520208-Additive-building-am-opportunities-in-a-digitalized-production.html. Accessed 10 Jan 2020
28. https://pdf.directindustry.com/pdf/phenix-systems/brochure-phenix-systems-gb/13610-346313.html. Accessed 07 Jan 2020
29. Fathi, M.H., Hanifi, A.: Sol-gel derived nanostructure hydroxyapatite powder and coating: aging time optimization. Adv. Appl. Ceram. **108**(6), 363–368 (2009)

Sintered Compacts of Co-Cr Powders Doped with HAp and ZrO_2 Used in Implantology

Diana-Irinel Băilă[1(✉)], Răzvan Păcurar[2], and Ancuța Păcurar[2]

[1] University Politehnica of Bucharest, Blv. Splaiul Independenței, no. 313, sector 6, 060042 Bucharest, Romania
baila_d@yahoo.com

[2] Faculty of Machine Building, Department of Manufacturing Engineering, Technical University of Cluj-Napoca, Blv. Muncii, no. 103–105, 400641 Cluj-Napoca, Romania

Abstract. Additive manufacturing (AM) methods are widely used in the industrial production, such processes being controlled by a computer, that permit to create three-dimensional object by depositing materials, usually in layers. The objective of this article was to realize sintered compacts of Co-Cr powder doped with ZrO_2 and with HAp, necessary to improve the bioactivity for the medical implants. For this study, the samples were immersed in simulated biological fluid (SBF) for 21 days. The samples were doped with different percentage of HAp, respectively ZrO_2. The morphology and the structure of the sintered compact samples were establish using a scanning electron microscope, realizing mapping analysis. In this study were remarked that the sintered samples of Co-Cr doped with ZrO_2, don't present the growth of HAp, but the sintered samples of Co-Cr present after immersion in SBF, an increase of new nucleons of HAp, especially for the sintered samples of Co-Cr doped with 10% HAp.

Keywords: Sintered compacts · Direct metal laser sintering · Co-Cr alloy · HAp · ZrO_2 · Bioactivity

1 Introduction

Additive manufacturing processes can be widely used for materializing of objects with precise geometric shapes, using computer aided design (CAD) or 3D object scanners. Direct Metal Laser Sintering (DMLS) process is built layer by layer, as with a 3D printing process, which contrasts with traditional manufacturing that often requires machining or other techniques to remove surplus material [1–15].

The alloy Co-Cr is a tolerable and non-toxic material for the human body and for this, it is used frequently for manufacturing implants, together with other metal alloys such as L316, L304 or Ti6Al4V.

Co-Cr alloys are metallic biomaterials which are essential to be used in orthopedy, cardio vascularity, and dentistry grace of their excellent mechanical properties, high corrosion resistance, and high wear resistance.

F. Gorski et al. (Eds.): MANUFACTURING 2022, LNME, pp. 69–78, 2022.
https://doi.org/10.1007/978-3-030-99769-4_6

Cobalt-chromium (Co-Cr) alloys have proper characteristics in terms of strength, hardness, toughness, corrosion and biocompatibility. Cobalt chromium, as well as titanium alloys are one of the most widely used materials for knee implants manufacturing. The mechanical properties of Co-Cr alloys are quite high and are similar to stainless steel material and the explanation for this is due to the multiphase structure and precipitation of carbides that allows the increasing of the hardness of Co-Cr alloys at the end [10–20].

The hardness of Co-Cr alloys is comprised between 550 and 800 MPa, and tensile strength is also variable between 145 and 270 MPa. If the Co-Cr alloys are heat-treated, the tensile and fatigue strength increases very much. These types of alloys present properties quite similar to the ones of the stainless steel material [20–36].

Generally speaking, the allergic reactions of the patients on the use of cobalt-chromium alloys are very low, as compared to the case of reported allergies of patients to specific metallic materials like nickel.

Zirconia is a bioactive material, a ceramic with excellent mechanical and thermal properties. In literature were realized thin film depositions of ZrO_2 on Stainless steel 316L lead to realize a monoclinic phase of ZrO2, and bioactivity in SBF [37, 38].

Hydroxyapatite is widely used material for dental restoration and orthopedic surgery, due to its chemical similarity to hard tissue.

Hydroxyapatite [HAp, $Ca_{10}(PO_4)_6(OH)_2$] is the most widely used calcium phosphate bioceramic materials in coatings of different metallic prostheses because of its osteogenic property and capacity of forming strong bonds with the host bone tissues. There are many methods available for realizing the HAp coating, as these are reported in the literature [39, 40].

2 Materials and Methods

In this research it has been used Co-Cr powder for the producing of parts made by direct metal laser sintering - DMLS process, with the following chemical composition: 54.31% Co; 23.08% Cr; 11,31% Mo, 7.85% W, 3.35% Si and Mn, Fe. The mechanical properties of the Co-Cr alloy powder are: Vickers hardness 375HV5, elastic module 229GPa, elongation at break 10%, elastic limit 0.2% 815 Mpa, corrosion resistance <4 $\mu g/cm^2$.

For the sintered samples made of Co-Cr by DMLS and doped with ZrO_2 afterwards there were used two types of percentage by mass 10% ZrO_2 and 20% ZrO_2.The raw materials powders were homogenized during an hour, using ethanol. The realized samples were pressed into the mold on an active section of 1 cm^2, using a pressing force of 20 kg/cm^2.

The hydroxyapatite used for doping; it was a powder with a purity greater that 90%. The sintered samples made of Co-Cr by DMLS process and doped with HAp, were realized in different percentage by mass (10%, 20% Hap) and the raw materials were homogenized 1 h, and the ethanol was used. The samples powders were pressed into mold in the same conditions as they were tested for the samples made of Co-Cr. and finally the pressed samples of Co-Cr doped with Hap and with ZrO_2 were sintered in an electrical furnace for post-processing.

The sintered compacts were immersed for 21 days in SBF1x (synthetic body fluid) at pH = 7.4, under sterile condition at the temperature of 37 °C.

After immersion in SBF, the sintered samples of Co-Cr alloy made by DMLS process and doped with ZrO_2 and HAp afterwards were studied using mapping analysis.

3 Experimental Part

The sintered compact samples of Co-Cr alloy doped with ZrO_2 and HAp, were studied before and after immersion in SBF, realizing electron microscopy and mapping analysis. The mapping realized before immersion in SBF in the case of sintered compact of Co-Cr doped with 10% HAp has showed a homogeneous mixed powder where the HAp has been detected as being present in composition in concentration of 10%, as shown in Fig. 1. A growth of HAp, existing in different zone with HAp agglomerations has been noticed after the immersion that has been performed in SBF (simulated biological fluid) in the case of sintered compacts of Co-Cr made by DMLS and doped with 10% HA afterwards, during 21 days of monitoring, as can be seen in the last image of Fig. 1 because of new HAp nucleons formation, can observe the peaks of calcium phosphate hydroxide Ca5(PO4)3(OH).

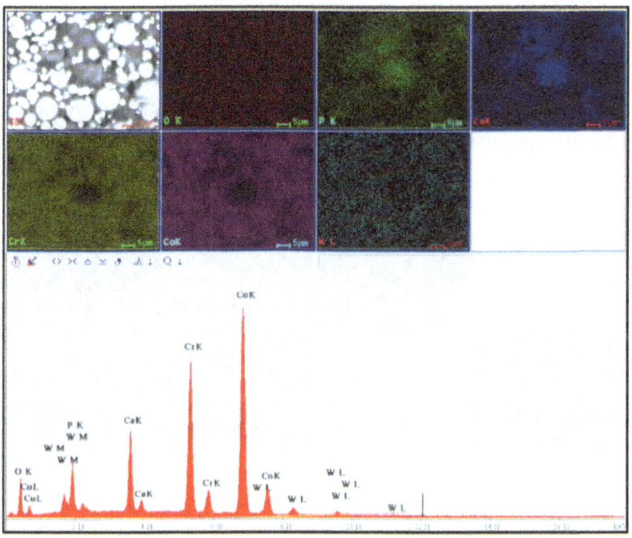

Fig. 1. Co-Cr doped with 10% HAp before immersion in SBF

In Fig. 2 is presented the mapping of sintered compact samples made of Co-Cr by DMLS and doped with 20% HAp, which is emphasizing the homogeneity of chemical elements. After immersion in SBF, the sintered compact made of Co-Cr alloy by DMLS and doped with 20% HA has been coated with a generous layer of hydroxyapatite and the new nucleons of HAp appear, having the grain of nanometer order, as in the last image of Fig. 2. Hydroxyapatite has irregular and elongated grains with nanometer size.

The mapping results presents the distribution of Ca and P zones, existing on the sintered samples.

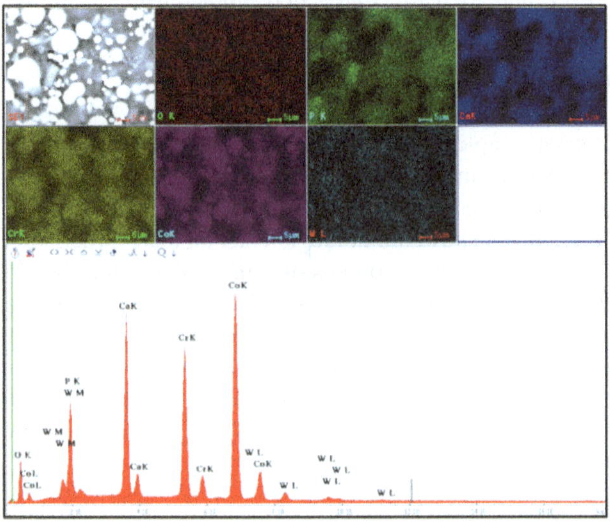

Fig. 2. Co-Cr with 20% HAp before immersion in SBF

In Fig. 3, it can be observed the mapping analysis of Co-Cr alloy powder doped with 10% ZrO_2, before immersion in SBF, for 21 days.

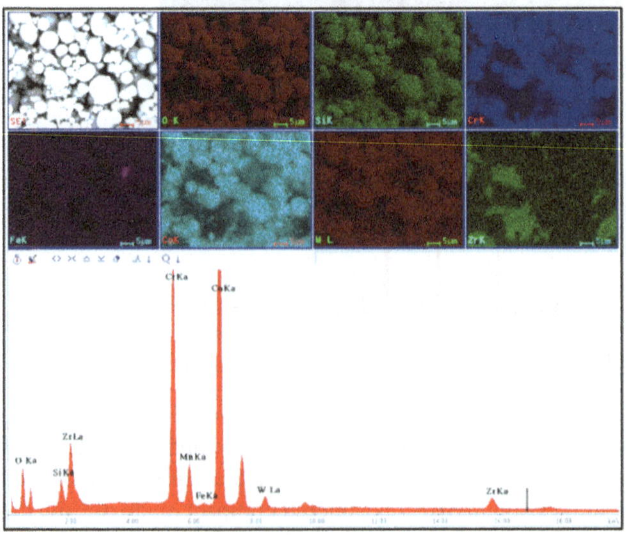

Fig. 3. Co-Cr with 10% ZrO_2 before immersion in SBF

Zirconia grains are spherical and are at size of nanometric scale after the immersion as shown in Fig. 4.

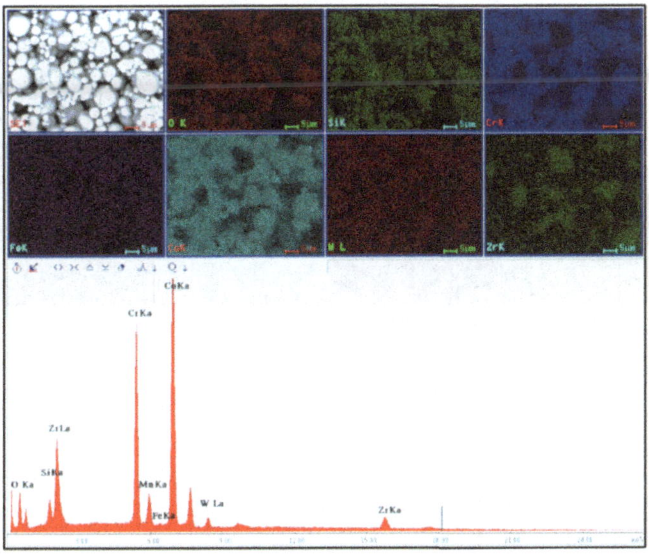

Fig. 4. Co-Cr doped with 20% ZrO_2 before immersion in SBF

After immersion in SBF, the morphology presents rounded form particles and news areas of zirconia crystallites of nanometer order also in the case of 10% Hap dopping, as shown in Fig. 5.

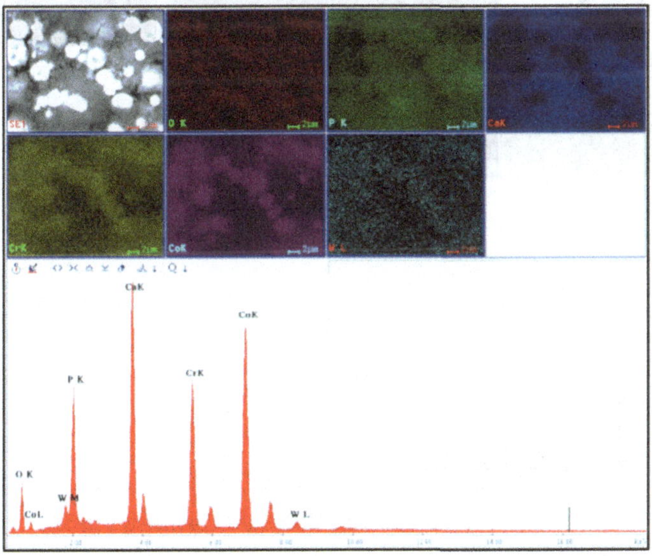

Fig. 5. Co-Cr doped with 10% HAp, after immersion in SBF

In Fig. 6, it is presented the mapping test of Co-Cr alloy powder doped with 20% ZrO_2, before immersion on SBF, for 21 days and can be noticed a greater concentration of ZrO_2 on the sintered sample in this variant.

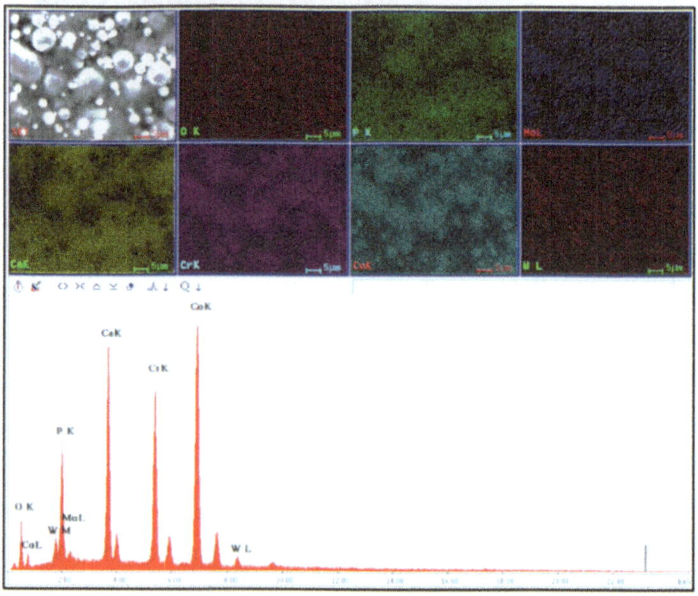

Fig. 6. Co-Cr doped with 20% HAp, after immersion in SBF

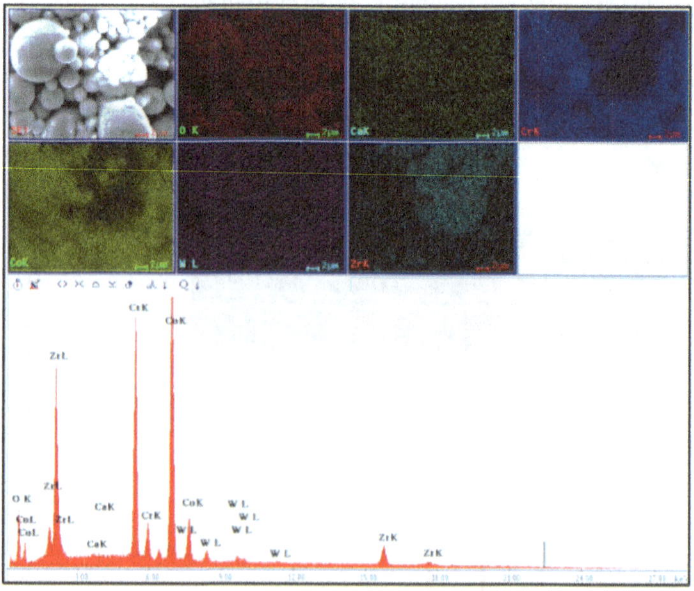

Fig. 7. Co-Cr doped with 10% ZrO_2, after immersion in SBF

In Fig. 7, the mapping analysis presents the uniform distribution of the chemical elements in the sample, the presence of ZrO_2.

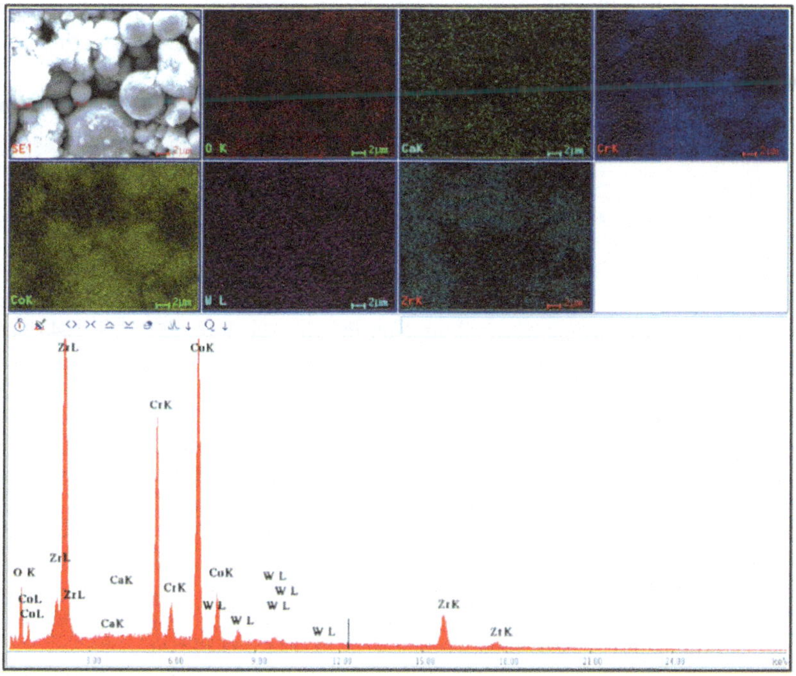

Fig. 8. Co-Cr doped with 20% ZrO_2, after immersion in SBF

Forming of new areas of zirconia crystallites of nanometer sizes after immersion in SBF (simulated biological fluid) of the sintered compacts made of Co-Cr by DMLS and doped with 10% HA afterwards has been noticed, during 21 days, as shown in Fig. 8.

4 Conclusions

The sintered samples that were obtained by DMLS process have emphasized good mechanical properties, but also fragile resistance in some areas due to the porous structure specific to sintered materials. In the case of orthopedic surgical applications that are realized based on hydroxyapatite composites, including of the primarily calcium and phosphate ion substances in the structure of the mixing material, as well as the controlling of the crystallinity and physical structures that ensures a better interaction of mixing elements for in vivo experiments are highly important since they are increasing in this way the biocompatibility and the possibility of using the reached surfaces of the mixing materials for different types of coatings that will further on enhance the opportunity for bonding the bones and the realized structures in the end.

In this article, for the sintered compacts made of Co-Cr by DMLS and doped with 10% ZrO_2, respectively 20% ZrO_2, after immersion in SBF, the appearance of new rounded fines particles of ZrO2 crystallites were remarked.

For the sintered compacts made of Co-Cr by DMLS and doped with 10% HAp and with 20% HAp respectively, after immersion in SBF, it was noticed the growth of hydroxyapatite concentrations, showed by the mapping analysis, aspect which is highly important since this will allow a better osseo-integration of the realized structures in the human body, by increasing the implants adhesion of the realized structures into the bone.

References

1. Mehrpouya, M., Dehghanghadikolaei, A., Fotovvati, B., Vosooghnia, A., Emamian, S.S., Gisario, A.: The potential of additive manufacturing in the smart factory industrial 4.0: a review. Appl. Sci. **9**, 3865 (2019)
2. Gu, D.D., Meiners, W., Wissenbach, K., Poprawe, R.: Laser additive manufacturing of metallic components: materials, processes and mechanisms. Int. Mater. Rev. **57**, 133–164 (2012)
3. DebRoy, T., et al.: Additive manufacturing of metallic components—process, structure and properties. Prog. Mater. Sci. **92**, 112–224 (2018)
4. Khaing, M.W., Fuh, J.Y.H., Lu, L.: (2001) Direct metal laser sintering for rapid tooling: processing and characterisation of EOS parts. J. Mater. Process Technol. **113**, 269–272 (2001)
5. Simchi, A.: Direct laser sintering of metal powders: mechanism, kinetics and microstructural features. Mater. Sci. Eng. A. **428**, 148–158 (2006)
6. Ford, S., Despeisse, M.: Additive manufacturing and sustainability: an exploratory study of the advantages and challenges. J. Clean. Prod. **137**, 1573–1587 (2016)
7. Lapcevic, A.R., Jevremovic, D.P., Puskar, T.M., Williams, R.J., Eggbeer, D.: Comparative analysis of structure and hardness of cast and direct metal laser sintering produced Co-Cr alloys used for dental devices. Rapid. Prototyp. J. **22**, 144–151 (2016)
8. Kellens, K., Renaldi, R., Dewulf, W., Kruth, J.P., Duflou, J.R.: Environmental impact modeling of selective laser sintering processes. Rapid Prototyp. J. **20**, 459–470 (2014)
9. American Center for Life Cycle Assessment. LCA XIV International Conference, San Francisco, CA, United States, pp. 130–141 (2014)
10. Diegel, O., Singamneni, S., Reay, S., Withell, A.: Tools for sustainable product design: additive manufacturing. J. Sustain. Dev. **3**(3), 68–75 (2010)
11. Su, X.S., Yang, Y.Q.: Research on track overlapping during selective laser melting of powders. J. Mater. Process. Technol. **212**, 2074–2079 (2012)
12. Dawes, J., Bowerman, R., Trepleton, R.: Introduction to the additive manufacturing powder metallurgy supply chain. Johnson Matthey Technol. Rev. **59**(3), 243–256 (2015)
13. Slotwinski, J.A., Garboczi, E.J., Stutzman, P.E., Ferraris, C.F., Watson, S.S., Peltz, M.A.: Characterization of metal powders used for additive building. J. Res. Natl. Inst. Stand. Technol. **119**, 460–494 (2014)
14. Tang, H.P., Qian, M., Liu, N., Zhang, X.Z., Yang, G.Y., Wang, J.: Effect of powder reuse times on additive building of Ti-6Al-4V by selective electron beam melting. JOM. **67**(3), 555–564 (2015)
15. Song, B., Dong, S.J., Deng, S.H., Liao, H.L., Coddet, C.: Microstructure and tensile properties of iron parts fabricated by selective laser melting. Opt. Laser Technol. **56**, 451–460 (2014)
16. Thijs, L., Verhaeghe, F., Craeghs, T., Van Humbeeck, J., Kruth, J.P.: A study of the micro structural evolution during selective laser melting of Ti-6Al-4V. Acta Mater. **58**, 3303–3312 (2010)
17. Băilă, D.I., Doicin, C.V., Cotruţ, C.M., Ulmeanu, M.E., Ghionea, I.G., Tarbă, C.I.: Sintering the beaks of the elevator manufactured by direct metal laser sintering (DMLS) process from Co-Cr alloy. Metalurgija **55**, 663–666 (2016)

18. Drstvensek, I., et al.: Applications of rapid prototyping in Cranio-maxilofacial surgery procedures. Int. J. Biol. Biomed. Eng. **1**, 29–38 (2018)
19. Gibson, I., et al.: The use of rapid prototyping to assist medical applications. Rapid Prototyp. J. **12**, 53–58 (2006)
20. Murra, L.E., et al.: Microstructure and mechanical behavior of Ti-6Al-4V produced by rapid-layer manufacturing, for biomedical applications. J. Mech. Behav. Biomed. Mater. **2**, 20–32 (2009)
21. Ciocca, L., Fantini, M., De Crescenzio, F., Corinaldesi, G., Scotti, R.: Direct metal laser sintering (DMLS) of a customized titanium mesh for prosthetically guided bone regeneration of atrophic maxillary arches. Med. Biol. Eng. Comput. **49**, 1347-1352. (2011). https://doi.org/10.1007/s11517-011-0813-4
22. Butscher, A., Bohner, M., Hofmann, S., Gauckler, L., Muller, R.: Structural and material approaches to bone tissue engineering in powder-based three dimensional printing. Acta. Biomater. **7**, 907–920 (2011)
23. Gibson, I., et al.: The use of rapid prototyping to assist medical applications. Rapid. Prototyp. J. **12**, 53–58 (2006)
24. Atzeni, E., Iuliano, L., Minetola, P., Salmi, A.: Proposal of an innovative benchmark for accuracy evaluation of dental crown manufacturing. Comput. Biol. Med. **42**, 548–555 (2012)
25. Pasebani, S., Ghayoor, M., Badwe, S., Irrinki, H., Atre, S.V.: Effects of atomizing media and post processing on mechanical properties of 17-4PH stainless steel manufactured via selective laser melting. Addit Manuf. **22**, 127–137 (2018)
26. Davis, J.R.: Nickel, cobalt and their alloys. In: ASM International, Materials Park, OH (2000)
27. Matkowic, T., Matkovic, P., Jadranka, M.: Effect of Ni and Mo on microstructure and some properties of Co-Cr dental alloys. J. Alloys. Compd. **366**, 293–297 (2004)
28. Meacock, C.G., Vilar, R.: Structure and properties of a biomedical Co-Cr-Mo alloy produced by laser powder microdeposition. J. Laser. Appl. **21**, 88–95 (2009)
29. Averyanova, M., Bertrand, P., Verquin, B.: Manufacture of Co-Cr dental crowns and bridges by selective laser melting technology. Virtual. Phys. Prototyp. **16**, 179–185 (2011)
30. Reclaru, L., Ardelean, L., Rusu, L., Sinescu, C.: Co–Cr material selection in prosthetic restoration: laser sintering technology. Solid. State. Phenom. **188**, 412–415 (2012)
31. Béreš, M., et al.: Mechanical and phase transformation behaviour of biomedical Co-Cr-Mo alloy fabricated by direct metal laser sintering. Mater. Sci. Eng. A. **714**, 36–42 (2018)
32. Barucca, G., et al.: Structural characterization of biomedical Co-Cr-Mo components produced by direct metal laser sintering. Mater. Sci. Eng. C. **48**, 263–269 (2015)
33. Girardin, E., et al.: Biomedical Co-Cr-Mo components produced by direct metal laser sintering. Mater. Today. **3**, 889–897 (2016)
34. Hunt, J.A., Callaghan, J.T., Sutcliffe, C.J., Morgan, R.H., Halford, B., Black, R.A.: The design and production of Co-Cr alloy implants with controlled surface topography by CAD–CAM method and their effects on osseointegration. Biomaterials **26**, 5890–5897 (2005)
35. Jacob, G., Brown, C., Donmez, A., Watson, S., Slotwinski, J.: Effects of Powder Recycling on Stainless Steel Powder and Built Material Properties in Metal Powder Bed Fusion Processes. NIST Advanced Building Series, pp. 100–106 (2016)
36. Building Readiness Level for ALM (Additive Laser Building) by Rolland Berger. https://docplayer.net/2520208-Additive-building-am-opportunities-in-a-digitalized-production.html. Accessed 10 Jan 2020
37. Thaveedeetraku, A., Witit-Anun, N., Boonamnuayvitaya, V.: Effect of sputtering power on in vitro bioactivity of zirconia thin films obtained by DC unbalanced magnetron sputtering. J. Chem. Eng. Jpn. **46**, 79–86 (2013). https://doi.org/10.1252/jcej.12we088
38. Denry, I., Kelly, J.R.: State of the art of zirconia for dental applications. J. Dent. Mater. **24**, 299–303 (2008). https://doi.org/10.1016/j.dental.2007.05.007

39. Awasthi, S., Pandey, S., Arunan, E., Srivastava, C.: A review on hydroxyapatite coatings for the biomedical applications: experimental and theoretical perspectives. J. Mater. Chem. B **9**, 228–249 (2021). https://doi.org/10.1039/D0TB02407D
40. Kim, H.W., Georgiou, G., Knowles, J.C., Koh, Y.H., Kim, H.E.: Calcium phosphates and glass composite coatings on zirconia for enhances biocompatibility. Biomaterials **25**(18), 4203–4213 (2004). https://doi.org/10.1016/j.biomaterials.2003.10.094

Contact Surface Model Parameterization of the Extra-Articular Distal Humerus Plate

Nikola Vitković[1(✉)], Miroslav Trajanović[1], Jovan Aranđelović[1], Răzvan Păcurar[2], and Cristina Borzan[2]

[1] Faculty of Mechanical Engineering in Niš, University of Niš, Niš, Serbia
{nikola.vitkovic,miroslav.trajanovic,
jovan.arandjelovic}@masfak.ni.ac.rs
[2] Faculty of Machine Building, Technical University of Cluj-Napoca, Cluj-Napoca, Romania
{razvan.pacurar,cristina.borzan}@tcm.utcluj.ro

Abstract. In orthopedic surgery, it is vital to use proper fixation techniques to treat various medical conditions. Distal humerus fractures include high energy trauma with ruptured skin and low energy trauma in osteoporotic bone. Surgical treatment of extra-articular distal humerus fractures using open reduction and internal fixation with plate implants improves patient recovery and decreases soft tissue complications. Nowadays, two types of plates and their variants are commonly used for bone fixation: Dynamic Compression Plates - DCP and Locking Compression Plates - LCP. For the extra-articular fractures, LCP Extra-articular Distal Humerus plates are generally used. These fixation systems are standardly anatomically shaped, and with the capability to provide angular stability to the bone. Depending on the bone shape and fracture properties (shape and position), it is not uncommon to perform additional plate bending during surgery, which can be a problem because of additional surgery time and the possibility of a wrong plate shape. To improve the topological and geometrical accuracy of the plate the Method of Anatomical Features (MAF) was applied and presented in this study. MAF enables the creation of a parametric plate-humerus contact surface model based on human humerus Referential Geometrical Entities (RGEs), Constitutive Geometrical Entities (CGEs) and Regions Of Interest (ROI). Using an anatomically shaped contact surface model, the standard plate model can be tailored to follow the shape of a particular humerus bone during pre-operation planning. The newly developed model can be manufactured using a specific manufacturing technology.

Keywords: Distal humerus · LCP plate · Parametric model · Method of anatomical features

1 Introduction

A prerequisite in orthopedic surgery is to deliver the finest medical therapy possible to a patient with bone trauma. Bone fractures are a common type of bone traumas, and they can be treated by using internal and external fixation techniques. The term external

F. Gorski et al. (Eds.): MANUFACTURING 2022, LNME, pp. 79–92, 2022.
https://doi.org/10.1007/978-3-030-99769-4_7

fixation refers to a surgical procedure where a fixator is placed outside the human body to stabilize bone fragments [1]. Internal fixation relates to the use of implants (screws, needles, and plates) to stabilize a bone fracture [2–5]. The use of internal fixation is more desirable because it enables a better recovery process [1, 2]. In the surgical treatment of bone fractures, the most used internal fixators are plates. These fixators are produced in different sizes and shapes so that they can be applied to specific patients [4]. Plate implants which are used for bone fracture fixation are standardized, and they are created by following standard recommendations and rules. This can raise a problem due to the difference between the shape and size of a patient's bone and a standardized plate implant surface [5–8]. When this happens, it can be difficult to find correct plate positioning, which can lead to inadequate load transfer during the healing of the bone, and patient recovery can be compromised. This issue can be reduced through the use of so-called personalized plate implants (PPI). PPIs shape and geometry are tailored to the morphology and anatomy of the patient's bone, and surgeon (clinical case) requirements, width possibility to preserve the blood flow in the periosteum [1, 8]. The application of PPI has a positive effect on patients, but it takes more time to complete pre-operative planning and implant production [8]. Consequently, in situations where the use of standard implants can cause intraoperative and postoperative complications, PPI is employed. It is generally known that a patient-specific bone model is required to develop an anatomically adjusted plate implant. In the cases of fractures, bone data acquired from medical imaging methods is usually incomplete, i.e., there is not enough geometrical or anatomical information to reconstruct a complete bone model. Therefore, in the research analysis, methods that enable the creation of complete models of bones based on incomplete medical data, obtained by medical scanning, will be presented. Secondly, details about plates and procedures (methods) for their creation will be described.

1.1 Creating Complete 3D Bone Models Using Incomplete Bone Data

It is common not to acquire complete 3D geometrical data of scanned bone in clinical practice. Besides the already stated bone fractures, other reasons for missing data can be insufficient 2D images obtained from X-ray, absence of volumetric scanning devices (e.g., Computed Tomography - CT), and inability to perform volumetric scanning (due to the extensive radiation exposure). In such cases, it is necessary to define and apply a method, which can enable the reconstruction of a complete 3D model of bone or a specific part of the bone, but only using the available data. The methods can be divided into methods based on a generic (template) model and those based on other techniques, mostly contoured projections [9]. Template model-based methods mainly use the Free Form Deformation (FFD) approach, statistical bone models, parametric bone models, or a combination of these models or strategies.

FFD methods are based on the use of a mesh model which is defined as boundary surface over a certain pre-defined volume model. The mesh model is adapted to the input model [9, 10] by using subsequent iterative deformations. Mesh can be deformed to accommodate a 2D X-ray of the bone in a particular projection or projections [11]. A similar technique (FFD) is a technique in which a previously created CT template model is deformed based on parameters read from X-rays [12].

Statistical bone models are based on the application of statistical methods on a set of input bone models acquired from different sources [13]. These models can be adapted to the patient's bone shape by applying specific values obtained from medical images. An example of creating a statistical-parametric model based on quadratic surfaces is given in [14]. This is a typical example of using statistics to form a model that can be modified depending on certain morphometric parameters (and even other parameters), obtained mainly from medical images. Statistically based models such as ASM (Active Shape Models) [15] are used to iteratively deform and fit the statistical model into the input model (e.g., X-ray). Another approach may be to apply a "smart" pattern model, where neural networks are trained with combined data from CT and X-ray, to predict the shape of the bone (radius and ulna) based on the patient's radiography [16].

Contour based methods use contoured projection in different directions, like application of 3D models based on anatomical characteristics and contours created over X-rays, as shown in [17].

1.2 Plate Implants

In modern medicine, different implant types are used to fix human bone fractures [18]. Oval-hole compression plates represent an introduction to the construction of Dynamic Compression Plates (DCP) [18]. Oval holes enable inter-fragment compression when inserted bolts are tightened. These holes are comparable to those described in [19] and are used for bone fragments compression during screw tightening. The advantages of DCP were the low incidence of improper jointing, stable internal fixation, and the absence of the need for external immobilization, which allows movement of adjacent joints. DCPs must be applied to the periosteum by pressing against the bone to ensure adequate stability and allow for bone functionality [1, 18–21]. Due to the resulting limited blood supply, this requirement poses one critical issue: the porosity of the cortical bone at the implantation location. However, some reservations have been raised [21, 22] about the usage of plates with a smaller contact area. Another issue with DCPs was refracture after plate removal from the patient's body. There is a recommendation that the plate shouldn't be removed for at least 15–18 months (while the gap between the bone fragments is eliminated) to avoid refracture [18]. Various researchers have looked at the causes of refracture and determined that it is caused by cortical necrosis [22, 23].

To lessen plate interference with cortical perfusion and consequently cortical porosity, a novel plate design was made. This novel design is referred to as the limited contact-dynamic compression plate (LC-DCP) [21]. LC-DCP has a lower surface-to-surface contact with the periosteum of the bone than DCP implants (about 50%), because of which cortical bone necrosis and osteoporosis beneath the bone are decreased. Furthermore, the LC-DCP is designed with plate-hole symmetry, allowing different-intensity dynamic compressions on both sides of the hole [21]. It's worth noting that certain researchers [23] have concluded that LC-DCP has no effect on bone blood flow or the biomechanics of the bone-implant assembly.

Almost all the aforementioned implants have now been replaced by plates that can perform locking as well as provide normal plate functions, such are Locking Compression Plates (LCP). Locked plating, on the other hand, cannot totally replace standard (conventional) plating [18]. It is possible to use a combination of both plating processes, and this should be done whenever is achievable [20–22]. In comparison to standard plates, LCPs provide better fixation and can sustain higher loads [25]. When locking screws are utilized, LCP does not require accurate contouring because the plate does not have to touch and/or align with a surface of the bone. In these cases, the plate serves primarily as a fixator, while keeping in mind that the wider space between the plate and the bone, can be problematic [24–26].

Distal humerus fractures of the human arm (elbow) represent 2–7% of all adult fractures [27]. Extra-articular fractures, as a particular type of distal humerus trauma, require open reduction and proper stabilization. Also, stable angular plates are required if the bone quality (e.g., affected by osteoporosis) is poor. Therefore, adequate elbow stabilization while the patient is in recovery is of great importance [27–29]. LCP (Locking Compression Plate) Extra-articular Distal Humerus plates are representatives of implants that can be used for this type of fracture because they enable anatomically shaped and angular stable fixation system [30]. To treat extra-articular fractures of the distal humerus, specific attention is focused on the LCP plate shape where the plate touches the distal part of the humerus, i.e., the distal plate end should curve along the back of the lateral column. Also, LCP distal plate part must be at safe distance from the *olecranon fossa* so that complete elbow extension is not impeded. To accomplish these requirements LCP plate needs to be bent during surgery in specific clinical cases [8]. Bending during surgery can be avoided if predefined PPI plates are used.

To better understand plate implants and their possible application to the human body, it is essential to understand some types of standard biomaterials which can be used for their production. The most common biomaterials utilized in medical practice for the manufacture of bone plates are bio-metals [31]. Bio-metals belong to the group of inorganic biomaterials. They are not biodegradable, although research is now underway to develop alloys that are. The bio-metals most commonly employed in orthopedic surgery include stainless steel, cobalt alloys, and titanium alloys. Because of corrosion, bio-metals can cause toxicity. Also due to heterogeneous load distribution in some cases, the application of bimetals is followed by early failure. Because of all of this, multiple studies [31–33] have tackled organic biomaterials and bio-composites, like Polyglycolic acid (PGA), Polylactic acid (PLA), Polymethyl-methacrylate (PMMA), Polydioxanone (PDS), and others. These studies proved that implants made of bio-composites and biopolymers can be used instead of metallic orthopedic implants. Today, they are only used in dental implants and in small flexible internal fixators.

1.3 The Proposed Solution for Plate Contact Surface Personalization

To improve the treatment of distal humerus fractures (preoperative, intraoperative, and postoperative procedures), specifically, extra-articular type, the authors propose the application of a modified PPI LCP plate model based on the standard LCP Extra-articular Distal Humerus plate developed with a new procedure presented in this paper. This procedure allows for the development of a geometric PPI LCP model whose contact surface with the bone and overall plate geometry are tailored to the geometry and morphology of the bone (or part of a bone), and parametrized. For this purpose, a PPI LCP parametric contact humerus-plate surface model was created using the Method of Anatomical Features (MAF) [6–8]. There are two possibilities for how this model can be used.

First, a standard LCP plate can be bent during pre-operative planning by using a physical template model produced by additive technologies and adjusted to patient bone shape. Second, PPI LCP plate CAD model can be used to produce a physical PPI LCP model (using additive or conventional production technologies) which can then be used as the plate implant during surgery.

2 Introduction to Method of Anatomical Features

MAF has already been applied to analyze and create different human long bone models [7, 8, 34, 35]. Geometrical models of organs can be created by using two general MAF procedures. The first procedure is based on Reverse Engineering techniques (RE), and the second uses generic models for the creation of patient-adapted models.

RE procedure is based on the application of medical imaging software for gathering and pre-processing data about the organ and CAD software used for additional processing. The procedure generally consists of the following steps:

- Medical imaging by using CT (for hard and soft tissue).
- Pre-processing (Segmentation) in medical software which enables creation of adequate organ models, e.g., hard or soft tissue. Segmentation is usually done by using the Hounsfield scale, named after Sir Godfrey Hounsfield. By using this scale different organs can be extracted from the radiographs.
- Exporting a 3D model from medical imaging software to an acceptable format. The usual export format is the STL format, widely used in RE, 3D printing and CAD.
- Importing and processing the 3D model in CAD software. The processing in CAD software is based on the following: additional filtering of unnecessary data, tessellation, creation of a polygonal model, creation of geometrical entities, and:

    - o Definition of the Referential Geometrical Entities (RGEs) – Entities (axes, planes, points, etc.) necessary to create all other geometrical features (e.g., curves, surfaces) on the human long bones.
    - o Definition of the Constitutive Geometrical Entities (CGEs) – Entities based on RGEs, which are used to develop geometrical models of higher-order (e.g., surface, solid)
    - o Construction of the human long bone models – Procedure for the creation of surface and solid models of human long bones.
    - o The definition and application of geometric Regions of Interest – They are based on anatomically defined ROI, but geometrically defined on the bone model.

In MAF, the parametric (generic) CAD bone model is defined as a point cloud model created based on geometric (RGEs, CGEs), morphometric and anatomical analysis of specific bone. Parameters are geometric and morphometric dimensions whose patient-specific values can be determined using medical imaging methods. By Using these values, the parametric model can be altered (tailored) to fit the morphology and anatomy of the human bone.

In this study, a parametric humerus-plate contact surface model (for PPI LCP) will be created based on the human humerus surface model built by MAF. This model will be defined using ROI, RGEs, CGEs, and custom geometrical elements defined during model building. The whole procedure for human humerus surface model creation, RGEs and CGEs is described in detail in [36], while in this study, only essential elements will be stated.

3 Creation of PPI LCP Parametric Model

Two distinct approaches to create a custom PPI LCP humerus-plate contact surface model based on a standard LCP extra-articular Distal Humerus plate will be shown (Fig. 1a).

- Curve Based Method (CBM) - Standard geometrical elements (spline segments) will be used as parametrized geometry, while the bone model will be used as a deformable template model (FFD).
- Subdivision Surface-Based Method (SDBM) will use ROI to define control mesh as the parameterized grid. Then, by modifying vertices and normals of the control mesh polygons, the SubD surface will conform to the bone template model.

Both methods will use an already created surface model of the left humerus [1, 36], obtained from a CT scan of a woman (50 years old), 512 × 512px, 0.5 mm scan thickness, scanned on Toshiba Aquilion 64-slice scanner, Clinical Center Nis. The applied CAD software was the DASAULT SYSTEM CATIA. The initial surface model of the humerus shaft and distal part is presented in Fig. 1b, with anatomically defined spline curves (CGEs) based on humerus RGEs. The simulated extra-articular fracture of the distal humerus (A2 type, OTA/AO classification) is created based on the descriptive model defined in the literature [37, 38] and shown in Fig. 1c.

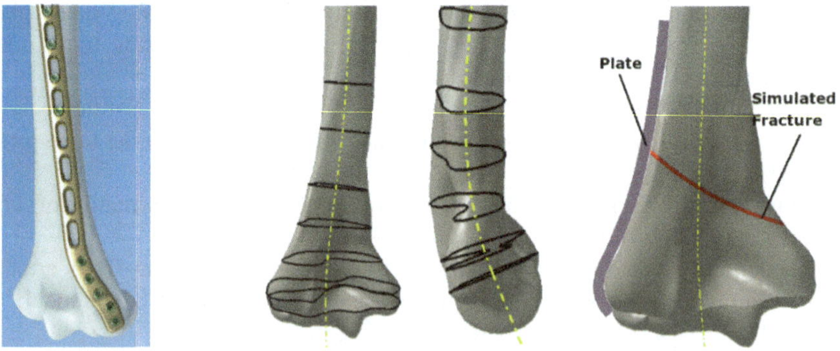

a b c

Fig. 1. Definition of LCP plate together with bone and fracture model; **a** – LCP plate model, **b** – Humerus surface model with anatomical curves (CGEs), **c** – Simulated fracture with LCP plate position

Both methods can be used to produce a newly developed PPI LCP model or use additive technologies to make a template model for bending a standard LCP plate.

3.1 Curve Based Method for PPI LCP Model Creation

For the development of a valid parametric model of the plate-humerus contact surface, it was essential to determine the position and orientation of the standard LCP plate. To perform this process, additional spline curves were created in distal humerus part and shaft, as presented in Fig. 2a. By using literature recommendation [30], the specific spline curves were created by trimming the existing curves (Fig. 2b). The guiding rules were to encompass the wider area in the humerus distal part and to make it more anatomically adjusted; to identify and avoid Olecranon fossa but get closer to it; to enable an extension of shaft part of the contact surface as much as possible. These curves conform to the guiding rules and enable the definition of a more complex and more detailed contact surface between plate and bone. The validity of these curves was checked by creating a surface model and checking it against the original (input from CT) polygonal model.

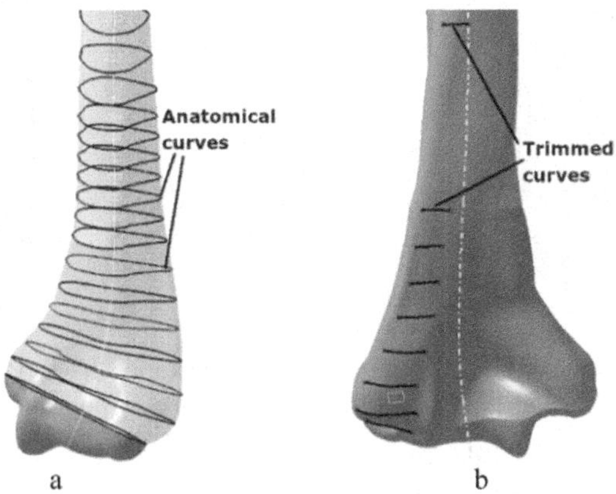

Fig. 2. Humerus surface and plate definition according to anatomical characteristics; **a** – Additional anatomical curves in distal and shaft part, **b** – Trimmed curves in the plate region.

The results are presented in Fig. 3a. Most deviations are below $\pm$ 0.15 mm which is more than acceptable. The created curves are applied to the definition of the surface model of plate contact area with humerus outer surface. The curves were used to create the PPI LCP contact surface model, and based on that model, the plate's volume (solid) model was created by adding thickness (Fig. 3b). Thickness is also a parameter value, and it can be adjusted to suit surgeon or literature recommendations (e.g., 2, 3, 4.5 mm).

The humerus surface model can be scaled and personalized by using X-ray images (Anterior-Posterior or Lateral-Medial view, or both views) of the patient bone (Fig. 3c). The spline curves will also be transformed and scaled to fit the patient's bone. Therefore, the humerus-plate contact surface model will be created according to the patient's bone anatomy and shape. If there is a need, curves can be additionally adjusted by transforming curve points manually. By adding thickness, PPI LCP plate model can be created and used as a template model created by additive technologies (e.g., Fused Filament Fabrication - FFF method) to bend standard LCP plate model, or it can be further processed by creating locking and regular holes and forming CAD model ready for Computer-Aided Manufacturing - CAM (i.e., milling, molding, forging, casting).

The created spline curves (Fig. 2b) are defined by using anatomical points. MAF anatomical points are parameterized by using morphometric parameters and linear regression to create parametric functions [7]. In the current research stage, the parametric model of the distal humerus part is at the testing phase, so the used points are already parametrized but not yet thoroughly tested. When this is completed, the humerus-plate contact surface model will be completely parametric, and it will be enough to create a personalized humerus bone (or distal part) model. Personalized PPI LCP plate model will be automatically created, as it is already built for the femur and tibia [7, 8, 34, 35].

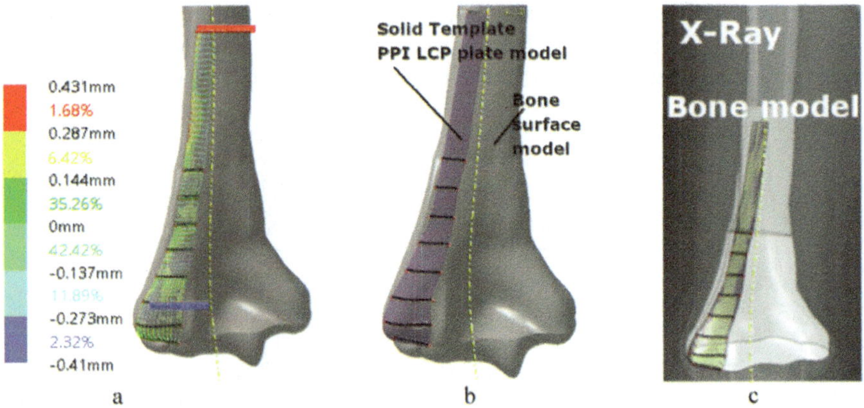

Fig. 3. Analysis of the created contact surface and implementation; **a** – Deviation analysis (bone – contact surface), **b** – Solid model of the plate template model, **c** – Application of the created model by using radiograph.

3.2 Subdivision Surface-Based Method for Creation of Parameterized Mesh

SubD objects are mesh-based, suitable for more approximate types of modeling such as character modeling and the creation of smooth organic forms. Parametric patches (surfaces) constructed over meshes of random topology are known as subdivision surfaces. In the same manner that a face or polygon forms a portion of the polygonal mesh, patches constitute "parts" of a larger area. Many distinct forms of patches have evolved to fulfill the objectives of geometric modeling. Furthermore, patches are made up of points or vertices that impact a rectangular area of a smooth surface [39]. Control points are the points that control the surface shape, and their set manages the patch as a control mesh or a control hull. What distinguishes different types of patches is how the control points impact their surface. Each control point in a mathematical sense has a "basic function" related to it, that affects the surface in a specific way when only that point is moved [39]. One of the simplest ways to create a set of patches is to apply a rectangular grid of control points. Since points can overlap in adjacent patches if one control point is moved, it affects multiple patches which ensures that they always meet seamlessly for B-spline patches. Bezier patches, on the other hand, share only points on their borders, which makes B-splines a better solution for surface representation.

For this study, parametric patches are defined across a humerus surface model and represented in Fig. 4a. First, patches are created in anatomical planes, which are defined as planes normal to the 3D anatomical axis. This way natural form of bone and SubD modelling principles are preserved. The next step is to define humerus-plate contact surface by limiting patches in the defined area, as it is presented in Fig. 4b. This approach enables forming of anatomically defined patches across the contact surface model. These patches are defined as custom geometric ROI, which are anatomically adjusted to the bone. Currently, eight patches, i.e., eight ROIs, are defined (more can be added). The nature of SubD surface modeling enables local control of the individual patches, but also, it can preserve C2 continuity across the whole shape of the humerus-plate contact surface. The individual control points of each patch are anatomical points defined by using RGES and CGES of the human humerus Eq. (1). These points are defined on the crucial anatomical and geometrical positions at the distal humerus and shaft. Using these control (anatomical) points, each patch can be manually modified (automatically if a parametric bone model is applied) and adjusted to the bone surface or surgeon requirements and join into one mesh Eq. (2) and (3). For example, it is possible to raise the plate contact surface, if there is a need for blood supply to the periosteum.

$$ROI_d = \{P_{dnm} | n = 1..rd, m = 1 \ldots xd\}$$

$$ROI_s = \{P_{snm} | n = 1..rs, m = 1 \ldots xs\} \tag{1}$$

$$ROI_{ds} = ROI_d \cup ROI_s \tag{2}$$

$$SubD_{rsd} = SubD(P_{dnm} \cup P_{snm}) \tag{3}$$

d – distal part; s – shaft part; r – radial points (along the anatomical axis); x – axial points (in planes normal/under specific angle to anatomical plane); ROI_{ds} – Union of two regions of interest, shaft and distal; $SubD_{rsd}$ – Meshed SubD surface which includes points from both ROI.

Two different SubD meshes are created by using CATIA net surface technical feature. The first mesh used two vertical guiding curves as control curves, and profile curves were automatically calculated, i.e., the mesh is automatically created based on the guiding curves limitations (Fig. 5a). The second mesh is created by using two vertical guiding curves, and horizontal profile curves (Fig. 5b). Only boundary patch curves are used, because the intention was to create a method that should be fast (optimal number of points or curves) and accurate enough (to satisfy clinical case requirements). If there is a need, additional points (additional curves) can be selected or created. Guiding and profile curves are the same ones used for the initial humerus-plate surface creation.

The deviation analysis between these surfaces (meshes) is presented in Fig. 5c, where it can be seen there is a difference at maximum of 1.87 mm, in the area where plate encompasses distal humerus. The next step was to compare individual deviation analysis between first and second mesh and bone surface created by using NURBS surface in CATIA. The analyzes are performed and presented in Fig. 6a, and 6b, respectively. First mesh has smaller deviations, but this is due to the nature of SubD creation, i.e., it has

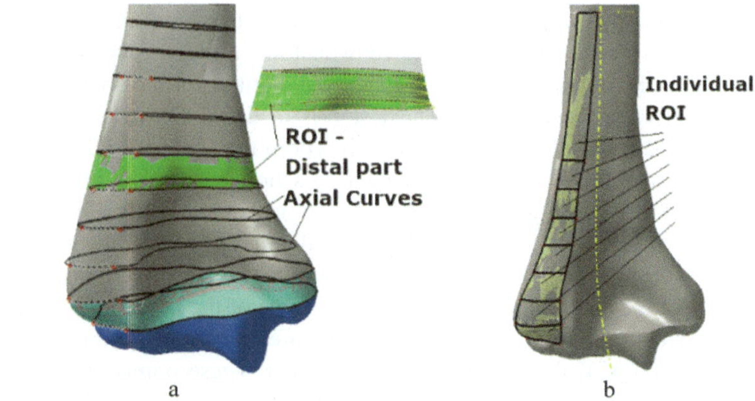

Fig. 4. ROI definition in specific regions and plate specific ROI; **a** – ROI examples (extracted one distal ROI), **b** – ROI defined for humerus-plate contact surface.

a continual surface flow. The second mesh has a more significant deviation error in the posterior-lateral part of the distal humerus. Deviations are around 1.85 mm (red dots). Again, this result can be excepted because it is limited by individual profile boundary curves and created as an individual patch concerning overall geometry. In both cases, the surface can be manipulated in selected areas and adjusted to the requirements using individual points, curves, patches, or meshes transformation.

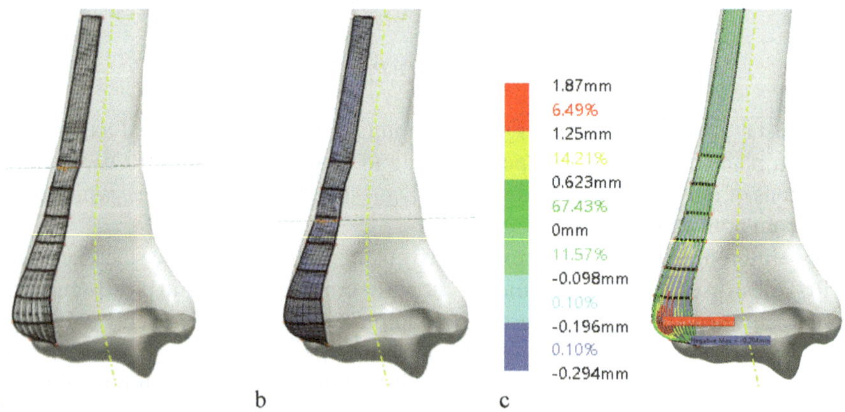

Fig. 5. SubD surface created by using ROI; **a** – Guiding curves mesh, **b** – Guiding and profile curves mesh, **c** – Deviation analysis between created SubD surfaces

In Fig. 6c, deviation analysis of the initial bone surface and modified second mesh surface is presented. In this case, the control mesh in the area with the largest deviation was adapted by removing just one boundary curve, and as it can be seen, the deviation error is below 1mm. Thus, with additional optimization, the geometrical accuracy of the model can be further improved, which conforms with the objective of this research, i.e., to demonstrate model geometrical (shape) adaptability to the patient's bone.

The whole process should be automated with the completion of the parametric humerus bone model. In that case, it should be easy to just change anatomical points coordinates by using the resulting personalized point cloud for specific human bone. SubD meshes should conform to the new position of control vertices, i.e., anatomical points, i.e., creating control mesh that conforms to the anatomy of the human bone.

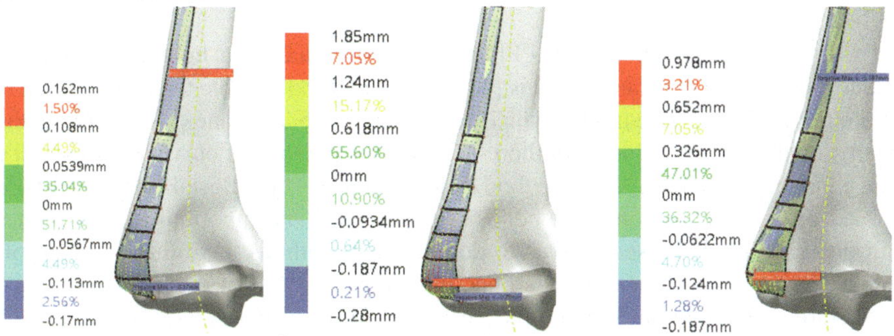

a b c

Fig. 6. Deviation analyzes ROI; **a** – First mesh and bone surface, **b** – Second mesh and bone surface, **c** – Deviation analysis between improved second mesh and bone surface

3.3 3D Printing of the Plate Model

Plate template and bone models were printed using CreatBot DX Plus 3D printer (FFF) as shown in Fig. 7a, 7b and 7c. Material was PLA, and production time was 13 h 21 min for bone, and 1 h 19 min for plate. The plate model was shaped to conform to the bone shape and anatomical characteristics. If there is a need to create metal implant, some other additive technology can be used, like DMLS (Direct Metal Laser Sintering)

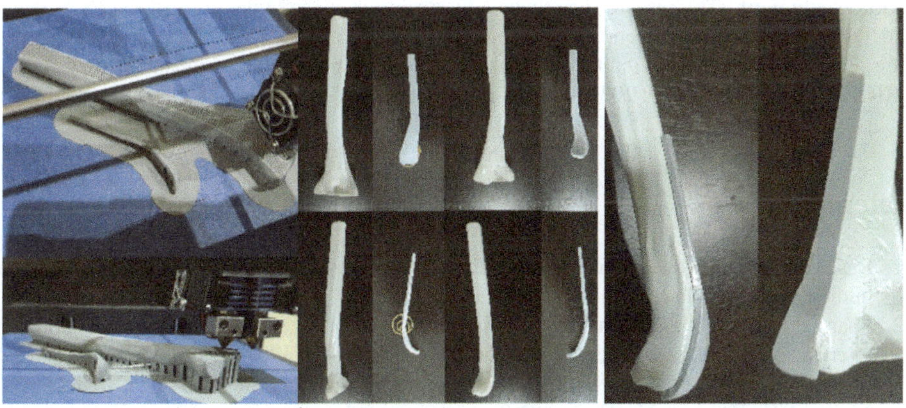

a b c

Fig. 7. Prototype production; **a** – 3D printing process, **b** – Postprocessed models, **c** – Bone and plate models assembly

which enables sintering of metal powdered by laser. This should allow personalization of plate implants and lead to their possible direct implantation into the patient body during surgery. In that way, pre-operative planning will be combined with surgical intervention, which will improve patient recovery.

4 Conclusion

Two methods for the creation of the PPI LCP Extra-articular distal humerus plate contact Surface model were introduced in this paper. Both methods enable the construction of a parametric contact surface model, which can be altered to fit a bone of a specific patient. The first method (Curve Based Method - CBM) uses parametric curves to construct the surface model, while the second method (Subdivision Surface-Based Method - SDBM) is based on the application of Subdivision surfaces. It is important to note that the main objective of this research was to demonstrate methods flexibility to adapt plate geometry and shape to the patient's bone and surgeon (clinical case) requirements in pre-operative planning, and to compare it to standardized plates and their application during surgery, particularly bending processes. In ideal circumstances, with proposed methods, we can fully align the plate model with the bone model and get no (with bone surface extracting method, thus making the plate contact surface the same as bone outer surface) or negligible deviations in geometry and shape, but as already stated that was not a primary goal. Therefore, it can be concluded that in the current stage of the research, both methods are applicable to adapt the PPI LCP plate model according to the requirements of the clinical case concerning bone geometry and anatomy. The future work should include one big improvement: the application of the parametric humerus bone model, which may be used for the creation of a patient-specific bone model. This addition to the method should provide complete and automatic adaptation of the contact surface model to the bone. Thus, it will enable easy modification of the PPI LCP plate model.

Acknowledgment. This research was financially supported by the Ministry of Education, Science and Technological Development of the Republic of Serbia, and ERASMUS+ project "Boosting the scientific excellence and innovation capacity of 3D printing methods in pandemic period" – BRIGHT. Project Reference: 2020-1-RO01-KA226-HE-095517.

References

1. Rashid, M., et al.: Towards patient specific plate implants for the human long bones: a distal humerus example. Facta Univ. Series: Mech. Eng. **16**, 347–357 (2018)
2. Sharma, S., John, R., Dhillon, M.S., Kishore, K.: Surgical approaches for open reduction and internal fixation of intra-articular distal humerus fractures in adults: a systematic review and meta-analysis. Injury **49**, 1381–1391 (2018)
3. Mulon, P.Y., Zarzosa, M., Harper, D.P., Anderson, D.E.: Assessment of two augmentation techniques on the mechanical properties of titanium cannulated bone screws. Am. J. Vet. Res. **81**, 116–121 (2020)

4. Moiduddin, K.: Implementation of computer-assisted design, analysis, and additive manufactured customized mandibular implants. J. Med. Biologic. Eng. **38**(5), 744–756 (2018). https://doi.org/10.1007/s40846-018-0370-5
5. Tomaževič, M., Kristan, A., Kamath, A.F., Cimerman, M.: 3D printing of implants for patient-specific acetabular fracture fixation: an experimental study. Eur. J. Trauma Emerg. Surg. **47**(5), 1297–1305 (2019). https://doi.org/10.1007/s00068-019-01241-y
6. Vitković, N., Milovanović, J., Korunović, N.: Software system for creation of human femur customized polygonal models. COMSIS – Comput. Sci. Inf. Syst. **10**(3), 1473–1497 (2013)
7. Majstorovic, M., Trajanovic, M., Vitkovic, N., Stojkovic, M.: Reverse engineering of human bones by using method of anatomical features. CIRP Ann. Manuf. Technol. **62**(1), 167–170 (2013)
8. Vitković, N., Mitković, M.M., Mitković, B.M., et al.: Reverse engineering of the mitkovic type internal fixator for lateral tibial plateau. Facta Univ. Series Mech. Eng. **13**(3), 259–268 (2015)
9. Filippi, S., Motyl, B., Bandera, C.: Analysis of existing methods for 3D modelling of femurs starting from two orthogonal images and development of a script commercial software package. Comput. Meth. Programs Biomed. **89**(1), 76–82 (2008)
10. Chenna, D., Yedukondala, N., Ghassemi, P., et al.: Free-form deformation approach for registration of visible and infrared facial images in fever screening. Sensors **18**(1), 125 (2018)
11. Ün, M.K., Avşar, E., Akçalı, İD.: An analytical method to create patient-specific deformed bone models using X-ray images and a healthy bone model. Comput. Biol. Med. **104**, 43–51 (2019)
12. Reyneke, C.J.F., Luthi, M., Burdin, V., Douglas, T.S., Vetter, T., Mutsvangwa, T.E.M.: Review of 2-D/3-D reconstruction using statistical shape and intensity models and x-ray image synthesis: toward a unified framework. IEEE Rev. Biomed. Eng. **12**, 269–286 (2019)
13. Schmutz, B., Rathnayaka, K., Albrecht, T.: Anatomical fitting of a plate shape directly derived from a 3D statistical bone model of the tibia. J. Clin. Orthop. Trauma **10**, S236–S241 (2019)
14. Sholukha, V., Chapman, T., Salvia, P.: Femur shape prediction by multiple regression based on quadric surface fitting. J Biomech **44**(4), 712–718 (2011)
15. Montúfar, J., Romero, M., Scougall-Vilchis, R.J.: Automatic 3-dimensional cephalometric landmarking based on active shape models in related projections. Am. J. Orthod. Dentofac. Orthop. **153**(3), 449–458 (2018)
16. Shiode, R., et al.: 2D–3D reconstruction of distal forearm bone from actual X-ray images of the wrist using convolutional neural networks. Sci. Rep. **11**(1), 1–12 (2021)
17. Gamage, P., Xie, SQ., Delmas, P., Xu, P.: 3D reconstruction of patient specific bone models from 2D radiographs for image guided orthopedic surgery. In: Proceedings of Digital Image Computing: Techniques and Applications, pp. 212–216. Australia (2009)
18. Lai, Y.C., Tarng, Y.W., Hsu, C.J., et al.: Comparison of Dynamic and Locked Compression Plates for Treating Midshaft Clavicle Fractures. Orthopedics **35**(5), 697–702 (2012)
19. Southeast Fracture Consortium.: LCP versus liss in the treatment of open and closed distal femur fractures: does it make a difference?. J. Orthop. Trauma **30**(6), e212–e216 (2016)
20. Milovanovic, J., Vitkovic, N., Stojkovic, M., Mitkovic, M.: Designing of patient-specific implant by using subdivision surface shaped on parametrized cloud of points. Teh Vjesn **28**, 801–809 (2021)
21. Berkin, C.R., Marshall, D.V.: Three-sided plate fixation for fractures of the tibial and femoral shafts: a follow-up note. J. Bone Joint Surg. Am. **54**(5), 1105–1113 (1972)
22. Perren, S.M., Cordey, J., Rahn, B.A., Gautier, E., Schneider, E.: Early temporary porosis of bone induced by internal fixation implants: a reaction to necrosis, not to stress protection? Clin. Orthop. **232**, 139–151 (1988)

23. Jain, R., Podworny, N., Hupel, T.M.: Influence of plate design on cortical bone perfusion and fracture healing in canine segmental tibial fractures. J. Orthop. Trauma **13**(3), 178–186 (1999)
24. Walsha, S., Reindla, R., Harveya, E., et al.: Biomechanical comparison of a unique locking plate versus a standard plate for internal fixation of proximal humerus fractures in a cadaveric model. Clin. Biomech. **21**(10), 1027–1031 (2006)
25. Rose, P.S., Adams, C.R., Torchia, M.E.: Locking plate fixation for proximal humeral fractures: initial results with a new implant. J. Shoulder Elbow Surg. **16**(2), 202–209 (2006)
26. Sanders, B.S., Bullington, A.B., McGillivary, G.R.: Biomechanical evaluation of locked plating in proximal humeral fractures. J. Shoulder Elbow Surg. **16**(2), 229–234 (2007)
27. Olson, J.J., Dyer, G.S.M.: Skinny wire and locking plate fixation for comminuted intra-articular distal humerus fractures: a technical trick and case series. JSES Rev. Rep. Tech. **1**, 34–40 (2021)
28. Antoni, M., Kempf, J.F., Clavert, P.: Comparison of bipolar and monopolar radial head prostheses in elbow fracture-dislocation. Orthop. Traumatol. Surg. Res. **106**(2), 311–317 (2020)
29. AO Foundation, Plate fixation and Open Reduction, https://www.aofoundation.org/. Accessed 20 Oct 2021
30. Scolaro, J.A., Hsu, J.E., Svach, D.J., Mehta, S.: Plate selection for fixation of extra-articular distal humerus fractures: a biomechanical comparison of three different implants. Injury **45**(12), 2040–2044 (2014)
31. Park, J., Kon Kim, Y.: Metallic biomaterials. In: Bronzino, J.D. (Ed), The Biomedical Engineering Handbook, CRC Press (2000)
32. Jahan, A., Edwards, L.K.: A state-of-the-art survey on the influence of normalization techniques in ranking: Improving the materials selection process in engineering design. Mater. Des. **1980–2015**(65), 335–342 (2015)
33. Hamdy, T.M.: Polymers and ceramics biomaterials in orthopedics and dentistry: a review article. Egypt. J. Chem. **61**, 723–730 (2018)
34. Vitkovic, N., Mladenovic, S., Trifunovic, M., et al.: Software framework for the creation and application of personalized bone and plate implant geometrical models. J. Health Eng. **2018**, 6025935 (2018)
35. Vitkovic, N., Radovic, L., et al.: 3D point cloud model of human bio form created by the application of geometric morphometrics and method of anatomical features: human tibia example. Filomat **33**(4), 1217–1225 (2019)
36. Rashid, M.M., Husain, K.N., Vitković, N., et al.: Geometrical model creation methods for human humerus bone and modified cloverleaf plate. J. Sci. Ind. Res. **76**(10), 631–639 (2017)
37. Scolaro, J.A., Voleti, P., Makani, A.: Surgical fixation of extra-articular distal humerus fractures with a posterolateral plate through a triceps-reflecting technique. J. Shoulder Elbow Surg. **23**(2), 251–257 (2014)
38. AO surgical reference, https://surgeryreference.aofoundation.org/orthopedic-trauma/adult-trauma/distal-humerus/extraarticular-simple-oblique/orif-lag-screw-with-neutralization-plate. Accessed 20 Oct 2021
39. Pharr, M., Jakob, W., Humphreys, G.: 03 – shapes, physically based rendering third edition. In: Pharr, M., Jakob, W., Humphreys, G., Kaufmann, M., (Eds.) pp. 123–244 (2016)

FEM Analysis of a Scoliosis Brace Concept with 3D Perforation for Manufacturing Using 3D Printing Technology

Natalia Różańska and Michał Rychlik(✉)

Institute of Applied Mechanics, Poznan University of Technology, pl. Marii Skłodowskiej-Curie 5, 60-965 Poznań, Poland
michal.rychlik@put.poznan.pl

Abstract. In recent years, many studies have focused on CAD/CAM systems for the design and manufacture of orthopaedic corsets for scoliosis. The rapidly growing field of 3D printing offers unique opportunities in biomedical applications, with benefits arising from the ability to customise the shape to the patient's body, and relatively low manufacturing costs. Currently, the goal is to minimise corset design in terms of both mass and volume of material used, while maintaining it's full corrective functionality. In addition, attention is also paid to aesthetic aspects. The use of 3D printing technology makes possible to develop an individually adjusted, lighter and more comfortable orthopaedic corset. This paper presents the results of FEM structural analyses carried out for two types of corsets: a standard full structure and a corset with 3D grid pattern structure. The analysis is a demonstration of the possibility of using pattern grid (many small holes) on the surface of the corset. Additionally, the influence of forces generated by the tension of fixing tapes on the corset model was examined.

Keywords: Brace · Scoliosis · Biomechanics · Finite element method analysis · 3D printing · 3D grid pattern

1 Introduction

An important role in the stabilization of the spine, and therefore the entire human body, is provided by an extensive ligamentous system and muscles. The function of ligaments is to keep the vertebrae in an appropriate position, enabling their physiological movement, which is safe for the spinal cord and nerves [10].

Scoliosis is a growth defect of the spine involving its deformation in three planes (frontal, sagittal, lateral). Scoliosis evolves mainly in the upper back in the thoracic segment (thoracic scoliosis) or between the thoracic and lumbar segments (lumbar scoliosis) [8, 12]. Idiopathic scoliosis is a three-dimensional spinal deformity (deformities occur simultaneously in all planes) that is most commonly diagnosed in children aged 10 years or older [5, 24]. The etiology of scoliosis development in many cases is unknown. Early detection of the spinal deformity makes it possible to prevent the progression of the deformity. Idiopathic scoliosis tends to worsen in developmental age during rapid growth of

F. Gorski et al. (Eds.): MANUFACTURING 2022, LNME, pp. 93–107, 2022.
https://doi.org/10.1007/978-3-030-99769-4_8

the spine. The risk of increasing the torsion angle increases as the degree of loss of physiological thoracic kyphosis increases [5, 12]. Therefore, early diagnosis of the disease and appropriate prevention are very important.

The treatment of scoliosis involves physiotherapy and kinesiotherapy, the use of an orthopedic corset and, in some cases, surgery. The treatment of idiopathic scoliosis with the use of a corset is a conservative treatment, which consists in stopping the progression of the curvature during the developmental period.

Treatment with an orthopedic corset is indicated for patients with an immature skeletal system, who have progression of torsion and have been diagnosed with a torsion of the spine with a Cobb angle of 25° [13]. On the other hand, contraindications for scoliosis treatment with an orthopedic corset include torsion exceeding the Cobb angle of 40°, pulmonary diseases, allergy to plastics, skin diseases in the back and trunk region [14].

The effectiveness of scoliosis treatment with an orthopedic corset depends on a number of factors, including [6, 7, 14]: correct manufacture and fit of the corset, compliance with the doctor's instructions on how to wear the corset, treatment carried out by a team consisting of a doctor, an orthotics' technician, a physiotherapist, and a psychologist, attendance at periodic outpatient and radiological check-ups, achievement of an initial torsion correction of approximately 50% from the initial value.

An orthopedic corset uses external forces designed to stretch, derotate and move the trunk to reduce spinal deformity in all three planes: frontal, sagittal and transverse. In some orthopedic corsets, cut-outs are used as pressure-relieving areas and, at the same time, reduce the weight of the corset and allow control of the patient's skin condition [15, 16, 18]. The mechanisms of action of a corrective corset can be divided into passive and active [15].

Various methods of manufacturing orthopedic corsets are currently in use, but with the development of computer-aided design and manufacturing, conventional methods of manufacturing orthopedic supplies are being replaced by CAD-CAM methods [22].

The procedure for making an orthopedic corset includes stages such as: consultation with the patient - including a clinical examination; taking measurements; individual measurement correction and corset design; fitting and acceptance of the final corset; follow-up visits [19].

The use of CAD-CAM methods significantly increases the comfort and speed of taking measurements for a corset for both the patient and the orthotics [22]. Currently, the operation of a system for assisting in the design of orthopedic supplies is based on three main elements [23]: digitization - 3D scanning; digital processing of the extracted data in CAD software; manufacturing of the positive orthosis.

Currently, corset manufacturing techniques are based on the positive body form of the patient. The corset structure is obtained by the thermoforming process [18, 22, 23]. Nevertheless, the manufacture of orthopedic corsets in the thermoforming process has some limitations. It is not possible to precisely control the thickness of the corset, nor is it possible to obtain complex textures and shapes, ensuring a high level of aesthetics, functionality and comfort.

An alternative method of manufacturing orthopedic corsets is incremental manufacturing technology, i.e. 3D printing. Three-dimensional printing allows obtaining objects with very complicated shapes, which would be difficult and very expensive to produce using another method. 3D printing using various methods [21, 23] significantly reduces the time and cost of manufacturing and testing of various products.

3D printing makes it possible to develop and manufacture a lighter and more comfortable orthopedic corset by adjusting the wall thickness according to individual patient features. The increased comfort in wearing the corset is due to the lighter weight, the presence of more flexible zones and better ventilation, while maintaining the corrective effect [9, 11, 25].

2 Orthopedic Corsets Manufactured with Standard and 3D Printing Technology Used in the Treatment of Scoliosis

2.1 Examples of Corsets Manufactured Using Standard Technology

Orthopedic corrective corsets are used to treat lateral scoliosis of the spine. Currently, there are several types of corsets on the market, the application of which determines the type of torsion of the spine. Traditional corsets differ in the way they press on the spine and the ribs to prevent curvature progression.

One of the first orthopedic corsets developed to treat scoliosis is the Milwaukee corset [18]. The Milwaukee corset is used for the treatment of curvatures of the whole length of the spine and when the correction of the thoracic and cervical spine is also necessary. This corset belongs to both corrective and relieving group of corsets. The spinal column is relieved by supporting the corset's structure on hip plates and supporting the mandible and occiput [13, 18]. Due to the impractical design of the corset and the need for adjustment to the patient, it is currently used only occasionally.

The Boston corset is used in the case of lumbar and thoracic curvatures, i.e. in the treatment of scoliosis with a curvature angle of 20–40°. The corset from the front covers the area of the body below the breast up to the pelvis, while from the back it covers the area below the shoulder blades to the tailbone [18]. The corrective effect is achieved by using holes in the corset to relieve the trunk on the opposite side from the applied load. The reduced dimensions of the corset improve the visual and cosmetic aspect and make the therapy more discreet [13, 18]. The Boston corset enables both passive and active correction. Thanks to this, it allows active work of muscles in the conditions of curvature correction and in this way protects them from atrophy.

The Cheneau corset is used in the treatment of thoracolumbar curvatures and for correcting vertebral rotation [13, 18]. The corset is recommended for patients when the curvature angle of the spine (Cobb angle) reaches values within 20–45° and at lower values when there is a significant progression of the curvature angle. The Cheneau corset is mainly manufactured from a polypropylene plate, which is shaped on a plaster model prepared from previously taken patient measurements. For Cheneau corset treatment to be effective, it is necessary to wear it most of the day and all night. It is only removed for corrective exercises, various hygienic activities and during other activities in which the corset would restrict movement and thus endanger the patient's safety [13, 18].

The PROVIDENCE corset is an orthosis designed for use at night when the patient lies on their back. It is usually combined in treatment with the Boston corset. The principle of operation of a night corset is based on hypercorrection of the patient's curvatures.

The main difference in the mechanism of action of this type of corset is that one arm of the corset is slightly raised. This results in the creation of lateral and rotational forces that correct spinal torsion [26]. The CEAN corset belongs to the group of corsets used only at night for the treatment of idiopathic scoliosis. It is manufactured from a thermoplastic material. This type of corset is mainly used when there are torsions between T5 and L3 and when the scoliosis is C-shaped. It can be worn alone or in addition to a corset worn during the day [26].

2.2 Examples of Corsets Manufactured Using 3D Printing Technology

The significant development of 3D printing during the last few years has resulted in the increasing use of this technology in the manufacture of orthopedic supplies.

3D printed corsets can achieve the same level of success in treating scoliosis as traditional corsets, while significantly improving wearing comfort.

The UNYQ Align corset is intended for the treatment of young idiopathic scoliosis in patients whose torsion angle does not exceed 40°. The construction of the corset is comfortable, light and breathable. These features are due to the use of 3D printing and perforated structure [2, 29]. Intel Curie sensors were placed in the corset, informing about how long the corset is worn by the patient. Additionally, the sensors monitor pressure points to ensure the best possible fit and treatment. A mobile app downloads the data from which adjustments to the design are made. This kind of connectivity helps patients to document their wearing time and therapy progress [2]. The patterns found on the corset allow ventilation of the skin, as well as giving it a fashionable appearance. This significantly improves the comfort and willingness of the patient to wear the corset [2, 29].

The Exos Armour corset has been developed to improve the quality of life of patients with spinal curvature [1, 17]. The corset uses 3D scanning technology to develop a perfectly fitting orthosis for the patient. Using the resulting model, it is possible to 3D print custom orthopedic corsets with high accuracy [1, 17]. The principle of the corset does not differ from the classic Cheneau and Boston corsets and the main advantage is the reduction of the production time of the corset from 20 days to 2–3 days, and maximization of the comfort of its wearing by the patient. The use of perforations makes it possible to reduce the weight of the corset by approximately 30% compared to a traditional corset.

The corsets made by WASP Medical [27] are a modified version of the Cheneau model. 3D printed corsets exclude the use of plaster moulds due to rapid CAD/CAM scanning and modeling. The use of these processes significantly reduces the cost of the orthosis and also ensures rapid delivery to the patient. The manufacturing process for this corset has been reduced, the complete preparation of the polypropylene corset takes ca. 30 h. This corset is characterized by a more anatomical shape than traditional corsets. This makes it more comfortable for the patient, so that the patient can wear it for a long period of time without feeling any discomfort.

3 Research Problem

The existing solutions of corsets made in standard technologies most often no longer meet contemporary requirements, both in medical and technological terms. The increased time needed to prepare a corset, the higher weight, as well as the lower degree of individual adjustment to the patient, result in reduced comfort and a longer treatment time [1, 13, 18, 26]. Corsets made with 3D printing technology have great potential, but the knowledge of their optimal shaping, adjusted to the type of scoliosis as well as personal needs of the patient, is still insufficient [3, 4]. This is particularly important in the aspect of making special 3D grid pattern structures (perforation) on the corset surface. These constructions allow the corset to be lighter, while at the same time ensuring adequate strength through optimal positioning. An additional advantage of the 3D grid pattern is the improved ventilation of the patient's skin, as well as the possibility of installing additional sensors [30]. Moreover, it is possible to manipulate the stiffness of the corset by appropriate shaping of the perforation shape and size, as well as their distribution (density of distribution of the set of holes). A new aspect that may be interested in the near future is the possibility to 3D print using different materials (filaments) with different mechanical properties, e.g. a material with higher stiffness or with increased flexibility.

Authors present a concept of an orthopedic corset made in 3D printing technology with the use of grid pattern structures (perforation). Moreover, the shape of the corset was adjusted with its usable parameters to individual anthropometric features of the patient. The aim of the conducted FEM analyses was to compare two corsets with a full-volume structure and with a 3D grid pattern structure. In addition, the effect of forces generated by the loads of the fixing straps on the corset was examined.

3.1 Preparation of 3D Computer Model of Corset with 3D Grid Pattern Structure

Designing a corset individually adjusted to the patient should begin with taking a 3D scan of the patient's body. On the basis of 3D scan and additional measurements, it is possible to obtain a model of the patient's body with appropriately assigned correction of body position. The patient model thus obtained is the starting surface for the creation of the orthopedic corset model in the CAD-CAM system (Fig. 1).

The torso model of the patient with spinal curvature was used to create the corset surface model (Fig. 1b). The corset design was made in Blender. The next step was to convert the STL model into an IGS file format. The converted file was processed in Inventor, where further operations were performed. Figure 1(c) shows the corset model with rectangular holes for fastening straps.

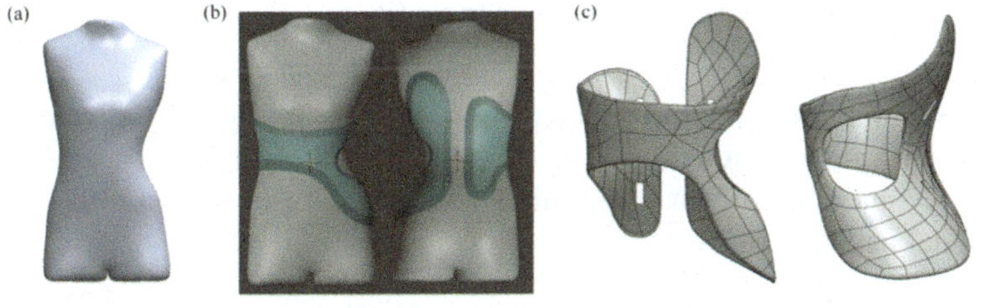

Fig. 1. Steps in modeling an orthopedic corset: (a) 3D scan of a patient showing curvature of the thoracic spine, (b) patient model with outlined corset area, (c) CAD model of corset (standard version with full structure) in Inventor software with places for fastening straps.

Creating a 3D grid pattern (perforation) on a surface model is possible in programs as Blender or Autodesk Meshmixer. These programs allow to obtain various types of 3D grid pattern structures (Fig. 2). The size of "holes" and their distribution is strongly dependent on the model grid. The use of this type of operation entails significant limitations. The whole surface of the corset is modified, without the possibility of limiting it to selected zones. When the perforation structure occupies the entire surface, it could significantly weaken the corset structure in terms of strength. For these reasons, it was finally decided to make the 3D grid pattern structure in Inventor.

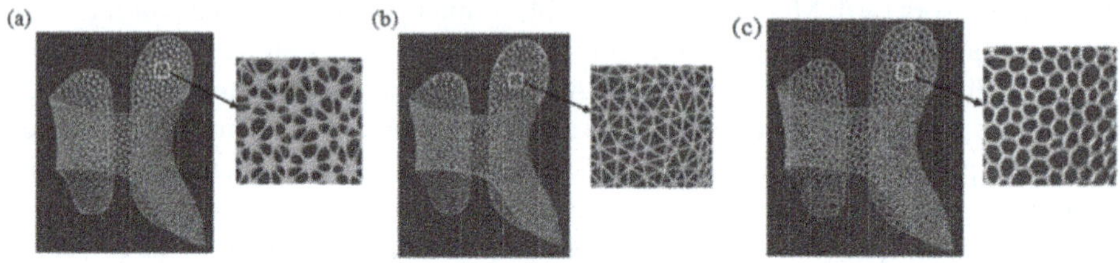

Fig. 2. Examples of 3D grid pattern holes structures created in Blender (a, b) and Autodesk Meshmixer (c) software.

In order to make the grid pattern, 3D sketches were developed, which were then projected onto the surface of the corset model. A triangle was chosen as the base element of the grid pattern structure. Next, on the basis of the sketches, a set of "holes" were made (3D grid pattern). In the last stage of work on the structure, the edges of the perforation were rounded (Fig. 3).

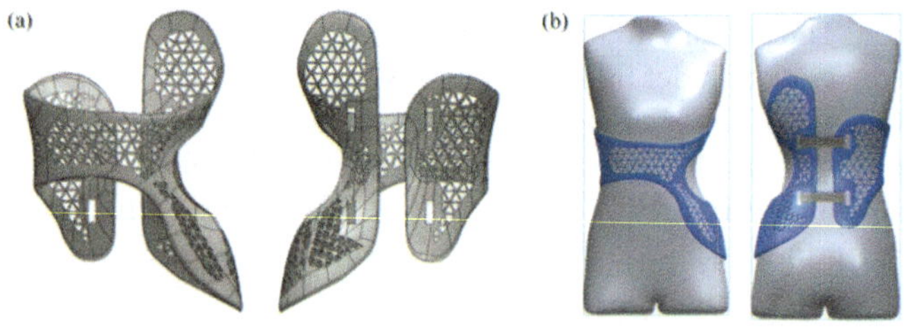

Fig. 3. CAD model of a corset with 3D grid pattern made in Autodesk Inventor: (a) 3D model of corset, (b) visualization of corset on patient's body model.

The preliminary design of the corset was based on the Rigo classification [20, 28] and a review of possible solutions. The Rigo classification describes spinal curvatures radiologically and clinically, and thus allows the shape of the corset to be adjusted to the individual patient [20, 28].

3.2 Preparation of Numerical Analysis of the Designed Corset – FEM

Due to the preliminary character of the research presented in this paper, the numerical analysis was carried out in the SolidWorks Simulation program. Two models of corset were prepared for analysis: Type 1 - a corset with a full structure and Type 2 - a corset with an applied 3D grid pattern. The purpose of using two types of models is to enable comparison of changes in strength properties of corset elements depending on the structure. A static testing will be chosen as the type of FEM analysis.

During the design of an orthopedic corset, it is very important to choose a material of sufficiently high strength so that the corset is not damaged and does not lose its corrective functions. In addition, some of the main requirements for a corset are that it be lightweight and waterproof.

As the corset is made using 3D printing technology, the materials considered for analysis were polylactic acid (PLA), ABS and polypropylene (PP). Finally, PLA was chosen for analysis in SolidWorks Simulation because it is a biodegradable polymer made from natural raw materials. After use, the material undergoes hydrolytic or biological degradation. Products made from this material have a smooth and glossy surface and, importantly, are easy to maintain, sterile and resistant to cleaning detergents and disinfectants [22]. Table 1 shows the mechanical properties of isotropic ABS, PP and PLA materials.

Table 1. Mechanical properties of PLA material used in the orthopedic corset.

Mechanical property	PLA
Young's Modulus [MPa]	3500
Yield Strength [MPa]	50
Poisson's ratio [-]	0.36
Density [$\frac{g}{cm^3}$]	1.25

In the next step, a finite element mesh was generated for the Type 1 corset in the high quality mesh had approximately 31k elements. The complex design of the corset Type 2 corset - with 3D grid pattern, generated numerous problems when creating the finite element mesh. For this purpose, the high quality mesh with the smallest element size was used. The total number of elements was over 44k elements.

The purpose of orthopedic corsets is to precisely place appropriate pressures and corrective forces on specific areas to minimize the progression of spinal curvature. Proper assessment of the pressures generated by corsets is an important factor in the effectiveness of treatment. In order to reproduce as real as possible the conditions between the corset and the torso, pressure fields were created on the inner surface of the corset. This pressure is the result of interaction between the inner walls of the corset and the human body.

The pressure fields and pressure values were developed on the basis of available literature [3, 4], in which simulations analyzing the effect of applied pressure on curvature correction were performed. Figure 4 presents a visualization of the external load boundary conditions, defined as pressure fields on the internal surface of the corset. The same pressure fields were used in simulations for both types of corsets.

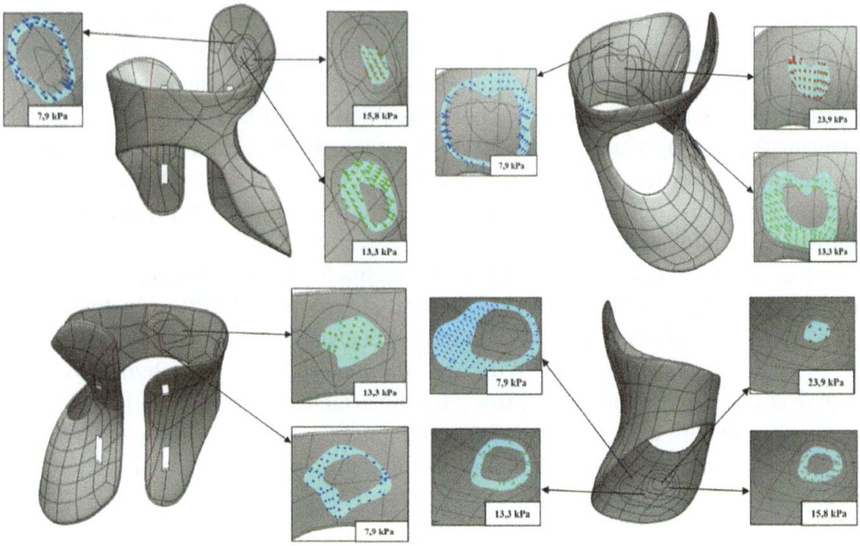

Fig. 4. Boundary conditions – external loads defined as pressure fields on the internal surface of the corset

In addition, the effect of strap tension on the resulting corset stress will be tested in simulations. The strap tension is not clearly defined, it is individually selected to suit the treatment of the individual patient (Fig. 5a).

For the numerical analyses, the corset models were fixed on the left side of the corset at the walls of the holes through which the fastening straps pass (Fig. 5b). In the next step, the forces acting in the horizontal direction were defined, reflecting the forces from the fixing straps. These forces were applied on the walls of the strap holes (Fig. 5c). A range of values from 20 N to 60 N has been identified as optimal in the literature [20, 28].

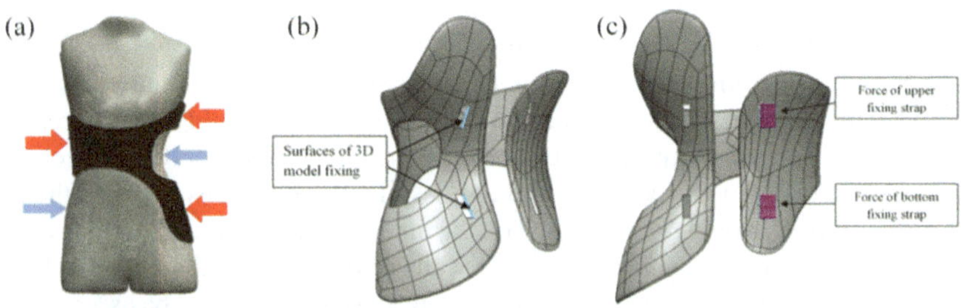

Fig. 5. Boundary conditions: (a) distribution of loading forces and pressure-relieving zones on the orthopedic corset, (b) 3D model fixing surfaces (marked in blue), (c) the points of application of the forces generated by the fastening straps.

4 Results of FEM Analysis of the 3D Printed Corset with Full Structure and 3D Grid Pattern Structure

4.1 Results of FEM Simulations for a 3D Printed Standard Corset Model with Full Structure

The first set of analyses were carried out for a standard orthopedic corset model made as a full structure. For the corset model, 4 simulations were performed assuming different tension values for the straps. In Simulations 1, 2, 3 the tension of the straps was defined as 20 N, 40 N, 60 N (equal values for the upper and lower strap), while in Simulation 4 the tension of the upper strap was defined as 60 N and the tension of the lower strap as 20 N.

The results of the FEM analyses in the form of stresses according to the von Misses hypothesis and displacement values are presented in Table 2 and Fig. 6.

Table 2. The results of the FEM analysis for the standard corset model with full structure.

	Tension forces on fixing straps	Max stresses von Misses [MPa]	Max displacements [mm]
Simulation 1	20 N (both straps)	34.4	78.1
Simulation 2	40 N (both straps)	35.4	53.4
Simulation 3	60 N (both straps)	41.4	88.2
Simulation 4	60 N upper strap 20 N bottom strap	31.2	44.3

In the case of Simulation 1, the reduced von Misses stresses predominate mainly at the front of the corset and in the vicinity of the relieving area, with values oscillating around 24 MPa. The maximum stress value occurs at the edge of the relief hole in the corset and reaches 34.4 MPa, not exceeding the PLA material strength of 50 MPa.

For Simulation 2, stress areas reaching values close to 24.7 MPa are mainly found in the area of the stress relief hole and the lower corset hole. A significant difference from Simulation 1 can be seen in the distribution and value of stresses in the front part of the corset. The highest stress is located in the corner of the lower strap hole on the inside of the corset and its value was 35.4 MPa.

In the case of Simulation 3, the stress distribution is very similar to previous Simulations 1 and 2, with a maximum stress value of 41.4 N. The stress values at the front of the corset to reach values in the range of 20–24 MPa.

In Simulation 4, the maximum stress was 31.2 MPa. This is the lowest value compared to the maximum stresses obtained for Simulations 1, 2 and 3. The stress distribution is comparable to the stress distributions obtained for tension forces of 20 N, 40 N and 60 N. In the case of Simulation 4, the stress accumulation is located around the relieving area and near the lower clamp hole. The stresses in these areas can reach values close to 21 MPa.

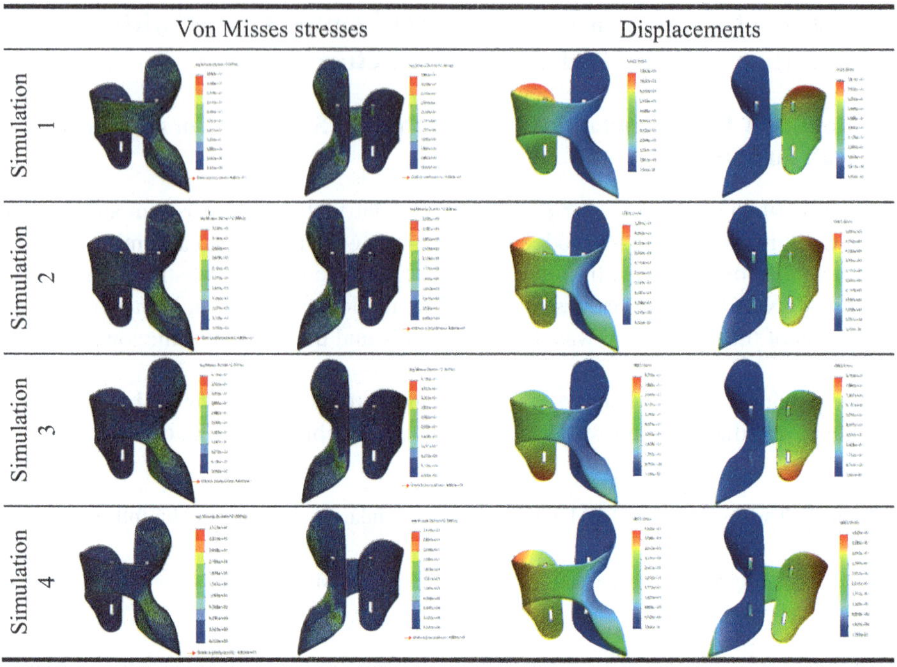

Fig. 6. Visualization of reduced stresses according to the Misses-Huber hypothesis and displacements for a standard corset with full structure – values in MPa and mm.

The maximum displacements for tensile strap forces of Simulation 1 (20 N), Simulation 2 (40 N) and Simulation 3 (60 N) were 78 mm, 53 mm and 88 mm. The lowest displacement occurs when a tension force of 40 N is applied (Simulation 2). Analyzing the results obtained, there is considerable variation in the location of the areas of the greatest displacement. The most desirable arrangement is a uniform and symmetrical distribution of displacements, which is observed for Simulation 2 (40 N force on the strap). For Simulations 1 and 2, it is apparent that the largest displacements occur at the top of the corset for a tension force of 20 N, while for a force of 60 N the largest displacements occur at the bottom of the corset. For Simulation 4, the maximum displacement was approximately 44.3 mm. The largest displacement can be observed in the lateral part of the corset in the direction opposite to the force.

4.2 Results of FEM Simulations for a 3D Printed Corset Model with 3D Grid Pattern Structure

A second set of analyses were carried out for an orthopedic corset model made with 3D grid structure, similar to the previous corset model. In Simulations 1', 2', 3' the tension of the straps was defined as 20 N, 40 N, 60 N (equal values for the upper and lower strap), while in Simulation 4 the tension of the upper strap was defined as 40N and the tension of the lower strap as 20 N.

The results of the FEM analyses in the form of stresses according to the von Misses hypothesis and displacement values are presented in Table 3 and Fig. 7.

Table 3. The results of the FEM analysis for the corset model with 3D pattern structure.

	Tension forces on fixing straps	Max stresses von Misses [MPa]	Max displacements [mm]
Simulation 1'	20 N (both straps)	30.2	75
Simulation 2'	40 N (both straps)	33.4	95
Simulation 3'	60 N (both straps)	35.6	156
Simulation 4'	40 N upper strap 20 N bottom strap	31.7	47.6

Analyzing the stress distribution for Simulation 1', there are several dominant areas where stresses can reach values of up to 30 MPa. The stresses accumulate in the anterior part of the corset on the left side and in the area of the relieving area.

In the next analysis, Simulation 2', the maximum stress occurs locally in the area of the lower band hole, near the model restraint. The stresses at these locations can reach values in the range of approximately 20 to 30 MPa. When comparing the results obtained for Simulation 1' and 2', it can be seen that in the case of the second one, the stress distribution in the front part has changed. On the left side, the stresses reach lower values than in the previous analysis.

The next results concern Simulation 3' in the front part of the corset, areas of stress accumulation are very visible. These can reach values in the range from 20 MPa to even 35 MPa. The analysis indicates areas especially exposed to damage under the influence of the tension of the fastening straps, and one of them is the area of constriction in the abdominal region.

The stress distribution in Simulation 4' is very uniform. The stresses mainly accumulate around the relief area and the holes made for the fixing straps. It should be noted that the stress distribution is more positive when two different forces are applied to the fixing straps (Simulation 4') with values of 20 N and 40 N. There are no areas of high stress accumulation and the stresses occurring in the corset reach safe values.

For each of the load cases of Simulations 1', 2' and 3', analyses of the distribution of the reduced displacements were carried out. The maximum displacements reached the values: 75 mm (Simulation 1'), 95 mm (Simulation 2'), 156 mm (Simulation 3') and they have a very similar pattern to the displacements obtained for the traditional corset (Simulations 1, 2 and 3). The largest displacements for the 20 N tension force (Simulation 1') are located in the upper left part of the corset from the fastening side (back). However, for the other two forces, 40 N (Simulation 2') and 60 N (Simulation 3'), the displacements with the largest values are located in the lower left part of the corset on the fastening (back) side. The displacement results obtained for Simulation 4' were also analyzed. The maximum displacement was 47.6 mm.

Analyzing the results of the simulations performed, it can be concluded that the use of different values of tension forces (upper and lower straps) can have a beneficial effect on obtaining a uniform distribution of forces on the corset, and thus also on the patient's body. The correct choice of strap tension force therefore has a significant effect on the corrective values of the corset, as it allows the corrective forces to be regulated.

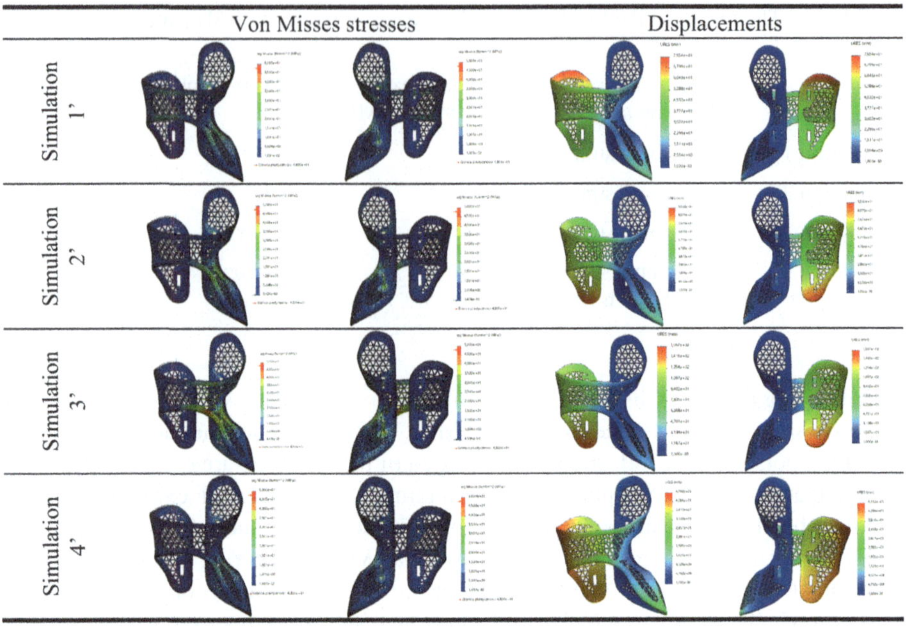

Fig. 7. Visualization of reduced stresses according to the Misses-Huber hypothesis and displacements for a corset with 3D pattern structure – values in MPa and mm.

5 Discussion

The FEM simulations carried out for two types of the corset make it possible to investigate the influence of the applied perforation on the change of loads occurring in the structure. The values of stresses occurring in the corset are very similar for both solutions. The maximum stresses occur locally and their location varies depending on the tension force of the straps used for testing. In the case of the corset with the perforated structure, the maximum stresses are significantly higher than the values obtained for the traditional corset. These stresses occur around the sharp edges of the holes, which may be due to the presence of a notch formed by the sharp corner of a triangular cell or an error in the FEM mesh. As these stresses occur at single nodes of the mesh, they do not have a significant effect on the strength of the entire corset structure.

The analysis of displacement visualization showed that in the case of applying equal strap tension forces both in the upper and lower part of the corset’s fastening, undesired asymmetrical displacements may appear. When a tension force of 20 N is applied in both analyzed corsets (Simulation 1 and 1’), the structure “opens” - in these cases the tension force of the fixing straps is too low. The results of displacement analysis for the standard corset with the tension force of 60 N (Simulation 3) do not exceed 90 mm. A similar displacement value was obtained for the corset with 3D grid pattern (Simulation 3’) using a lower tension force of 40 N. Accordingly, for a tension force of 60 N, the corset with 3D grid pattern structure (Simulation No. 3’) obtained maximum displacements almost twice as high (156 mm) compared to the displacements in the traditional corset

(Simulation 3). The reason for this phenomenon may be that the total internal surface area of the corset with the cellular structure, on which the pressure forces coming from the body act, is smaller than that of the traditional corset. This is due to the presence of numerous holes in the surface of the corset.

6 Conclusions

A characteristic feature of 3D printed corsets is the ability to make a 3d grid pattern structure. Making this type of hole structure is much simpler using additive techniques than during, e.g. thermoforming. Moreover, with 3D printing it is possible to obtain complex shapes of the corset, which is often impossible to obtain with traditional manufacturing techniques. A very important issue is the proper design of the structure of cutouts (3D grid pattern), especially as regards the shape, size of elements and their distribution among each other. The analyses carried out indicate places which may be especially exposed to stresses under cyclical loads. Moreover, the corset is subjected to significantly different forces during putting on and taking off than during use. The numerical simulations carried out give a rather limited view of the real conditions occurring during 24-h use of the corset.

The obtained results of the analysis are illustrative and basic. Further work should be considered to determine the most suitable boundary conditions as well as material conditions (use of anisotropic material model better reflecting the properties of the 3D printed material). The quality of printed parts and their mechanical properties depend on the parameters of the manufacturing process like infill density, extrusion temperature or layer thickness. The conducted analysis did not take into account the characteristics of the material, which are influenced by the parameters of the manufacturing process. Further research should consider influence of manufacturing method and process parameters on obtained mechanical properties of the orthosis.

Moreover, additional research would have to be conducted on the optimization of the designed corset structure - in particular, the search for the optimal shape of 3D grid pattern and their distribution on the corset. The above-mentioned potential directions of further research on the structure of a corset with a 3D grid pattern do not cover all the possibilities of analyses to which such a corset should be subjected, but only show certain directions. These analyses would certainly provide many important results influencing the overall development of the design shape in search of the final and most optimal solution for the orthopedic corset.

References

1. Canavese, F., Kaelin, A.: Adolescent idiopathic scoliosis: indications and efficacy of nonoperative treatment. Indian J. Orthop. **45**(1), 7–14 (2011). https://doi.org/10.4103/0019-5413.73655
2. Chung, C., et al.: Mechanical testing of a novel fastening device to improve scoliosis bracing biomechanics for treating adolescent idiopathic scoliosis. BioRobot. Lab. Appl. Bionics Biomech. **2018**, 10 p (2018). https://doi.org/10.1155/2018/7813960. Article ID 7813960

3. Clin, J., Aubin, C.E., Parent, S., Labelle, H.: Biomechanical modeling of brace treatment of scoliosis: effects of gravitational loads. Med. Biol. Eng. Comput. **49**(7), 743–753 (2011). https://doi.org/10.1007/s11517-011-0737-z. PMID 21287287
4. Cobetto, N., et al.: Braces optimized with computer-assisted design and simulations are lighter, more comfortable, and more efficient than plaster-cast braces for the treatment of adolescent idiopathic scoliosis. Spine Deform. **2**(4), 276–284 (2014). https://doi.org/10.1016/j.jspd.2014.03.005
5. Dickson, R., Lawton, J., Archer, I., Butt, W.P.: The pathogenesis of idiopathic scoliosis. Biplanar Spinal Asymm. Bone Joint Surg. **66**(1), 8–15 (1984). https://doi.org/10.1302/0301-620X.66B1.6693483
6. Głowacki, M., Kotwicki, T., Kaczmarczyk, J.: Curvatures of the spine. In: Dega, W. Kruczyński, J., Szulc, A., PZWL, Warsaw Orthopaedics and Rehabilitation, under red, pp. 421–443 (2015). ISBN 978-83-200-4961-9
7. Grivas, T., Mauroy, J., Wood, G., Rigo, M.: Brace classification study group (BCSG): part one – definitions and atlas. Scoliosis Spinal Disorders **11**(1), 1–47 (2016). https://doi.org/10.1186/s13013-016-0102-y
8. Gzik, M.: Biomechanics of the Human Spine Silesian University of Technology Publishing, Gliwice (2007). ISBN 978-83-7335-449-4
9. Gzik, M., Gąsiorek, D.: Application of Additive Technologies in Bioengineering, Faculty of Biomedical Engineering, Silesian University of Technology
10. Ignasiak, Z.: Anatomy of the Musculoskeletal System. Edra Urban & Partner, Wrocław, 2 (2020). ISBN 978-83-7609-945-3
11. Jin, H., et al.: Case series: 3D printed orthopedic brace combined with traditional manipulative physiotherapy to treat new-onset scoliosis in adults. Medicine **101**(1), e28429 (2022). https://doi.org/10.1097/MD.0000000000028429
12. Karpiński, M., Kamińska, M.: Idiopathic scoliosis, postgraduate paediatrics. UDSK Bialystok **15**(4), 75–79 (2011)
13. Kijowski, S.: From the Ambroise Pare armourer's corset to the dynamic corset - rehabilitation supplies for the treatment of scoliosis. Med. Rev. Univ. Rzeszów. Rzeszow 2009, 4, 418–423 (2009). ISSN 1730-3524
14. Korbel, K.: Results of Corset Treatment of Idiopathic Scoliosis According to Guidelines Scoliosis Research Society. Ph.D. Study. Department of Paediatric Orthopaedics and Traumatology Department of Spine and Orthopaedics, Poznan (2017)
15. Kotwicki, T.: Idiopathic Scoliosis - Pathology, Pathology, Diagnosis, Management, Ortopedia, Traumatology Rehabilitation Medsportpress (2013)
16. Kotwicki, T., Dumała, J., Czaprowski, D.: Principles of non-operative treatment of idiopathic scoliosis - guidelines based on SOSORT recommendations (Society on Scoliosis Orthopaedic and Rehabilitation). Orthop. Traumatol. Rehabil. **5**(6), Vol. 11, 379–395 (2006)
17. Leda, H.: Engineering Materials in Biomedical Applications, Poznan University of Technology Edition (2012). ISBN 9788377751114
18. Przeździak, B.: Rehabilitation Supplies, Via Medica – Medical Publishing Gdansk (2003). ISBN 83-89493-81-0
19. Redaelli, D.F., et al.: 3D printing orthopedic scoliosis braces: a test comparing FDM with thermoforming. Int. J. Adv. Manuf. Technol. **111**(5–6), 1707–1720 (2020). https://doi.org/10.1007/s00170-020-06181-1
20. Rigo, M.D., Villagrasa, M., Gallo, D.: A specific scoliosis classification correlating with brace treatment: description and reliability. Scoliosis **5**, 1 (2010). https://doi.org/10.1186/1748-7161-5-1
21. Siemiński, P., Budzik, G.: Additive Technologies: 3D Printer Printing. Publishing of the Warsaw University of Technology (2015). ISBN 978-83-7814-255-3

22. Zenkiewicz, M., Szach, A.: Selected problems in thermoforming of polymeric materials. Polymers **LV**(5), 335–418 (2010)
23. Wyleżoł, M., et al.: Biomedical Engineering - Incremental Methods in Medical Technology, Lublin University of Technology Lublin (2016). ISBN 978-83-7947-211-6
24. Weiss, H.R., et al.: Indications for conservative management of scoliosis. Scoliosis **1**, 5 (2006). https://doi.org/10.1186/1748-7161-1-5
25. Weiss, H.R., Tournavitis, N., Nan, X., Borysov, M., Paul, L.: Workflow of CAD/CAM scoliosis brace adjustment in preparation using 3D printing. Open Med. Inform. J. **11**, 44–51 (2017). https://doi.org/10.2174/1874431101711010044
26. https://bespoke.info.pl/category/gorsety-ortopedyczne/. Accessed 21 Aug 2021
27. https://fabbaloo.com/wp-content/uploads/2020/05/image-asset_img_5eb0bb9bd41c8.jpg. Accessed 21 Aug 2021
28. https://www.ortopedialopez.com/tienda/corses/corse-de-milwaukee/. Accessed 21 Aug 2021
29. https://www.skolioza.com/gorset.htm. Accessed 22 Nov 2021
30. https://vustudents.ning.com/profiles/blogs/3d-printed-osteoid-cast-uses-ultrasound-to-speed-up-bone-healing?xg_source=activity. Accessed 22 Nov 2021

Mechanical Properties of a Novel Device for Treatment of Neuropathy

Agata Kuliberda(✉) and Tomasz Stręk

Institute of Applied Mechanics, Poznan University of Technology, Poznan, Poland
agata.kuliberda@student.put.poznan.pl,
tomasz.strek@put.poznan.pl

Abstract. This paper presents an analysis of the mechanical properties of the novel device, which is used in the treatment of neuropathy. According to the human anatomy, a model of this device was designed taking into account the characteristic dimensions. The survey was conducted using Solid Works. The results of the conducted research provide information about the structural strength of the device under a given load. The values of stresses and displacements were determined by finite element simulation.

Keywords: Neuropathy · Finite element method · Rehabilitation device

1 Introduction

Transmission of nerve impulses to individual organs is carried out by nerve fibers of the peripheral nervous system. However, these fibers can be damaged by mechanical means or by chronic disease. The phenomenon of nerve fiber damage is called neuropathy. This disorder can affect only one fiber - this is mononeuropathy or multiple fibers – polyneuropathy. Some of the most common symptoms that patients present to the doctor with are muscle weakness, sensory disturbances, and sweating disorders. There are also changes visible on the skin [1–3].

Both the patient and the doctor must decide which neuropathy treatment will be most appropriate for the condition. The solution may be not only pharmacotherapy but also kinesitherapy. Most often, both types of therapy are used simultaneously.

Rehabilitation equipment available in the market can be either portable or stationary. They help in achieving maximum stability in the joints as well as strengthening the damaged ligaments and tendons. With these solutions, restoring the proper biomechanics of damaged joints is faster and more effective. A small group of equipment is used for the rehabilitation of patients with nervous system disorders. Rehabilitation devices may be equipped with a motor, but it is not necessary. Some of them require the activity and involvement of the patient, and in working with some of them the patient does not have to make any special movements, but only adopts a proper position [4].

For example, rehabilitation of the ankle joint can involve a robot that works using the Stewart platform principle. It is a parallel mechanism. The robot has 2 platforms, one of

F. Gorski et al. (Eds.): MANUFACTURING 2022, LNME, pp. 108–119, 2022.
https://doi.org/10.1007/978-3-030-99769-4_9

which is movable and the other is fixed. The moving one is a three-link serial manipulator with three rotational DOFs. Equipped with rotary encoders, it is possible to measure real-time angles of motion at the ankle joint, record torques and other human-robot interactions [5].

Electrotherapy has become popular in the treatment of neuropathy. Electrical stimulation (TENS) triggers the local release of neurotransmitters such as serotonin and relieves pain in the patient's limb. This method works well for diabetic neuropathy [6].

Unfortunately, there is still a lack of rehabilitation equipment to assist in the treatment of the other neuropathy types. A solution is needed to replace the physical therapist in manual therapy. The patient should be able to safely use such a device in their home because in many cases they need to be accompanied by a caregiver [7].

This paper presents the concept of a device that could be used during the rehabilitation of patients suffering from different types of neuropathy, regardless of the nature of their origin. Thanks to this device it will be possible to reduce the necessity of active participation of physiotherapists in treatments. Additionally, the patients will be able to have the device at their residence, making it easier for their family to care for them. This is important because patients with balance disorders are not able to get to a specific rehabilitation center on their own, and many people cannot afford frequent home visits from a specialist. The device could also be used in physiotherapy offices or rehabilitation centers. This will allow the doctor or physiotherapist to see more patients at the same time.

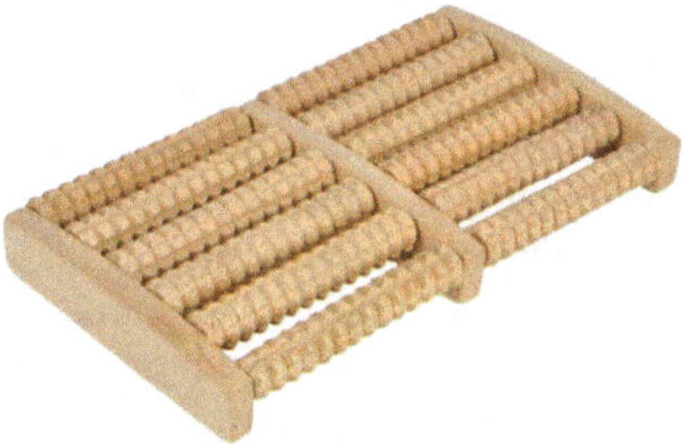

Fig. 1. A wooden foot massage roller that requires patient or physical therapist involvement.

The device that inspired the design presented in this article is a wooden structure without a drive (Fig. 1). The wooden foot massager affects the reflexes that are located on the feet, so it improves the condition of the body to a great extent. The movable rolls gently massage the whole foot, thus inducing a state of relaxation and providing relief from neuropathy. The therapy affects vasodilation, metabolism, physical and mental performance. The device should be placed on a stable surface. Place your feet on the massager, press your feet against the toothed wheels with moderate force while making a reciprocating motion. The disadvantage of this device is that the patient or physical therapist must be involved in the therapy. The patient must perform forward and backward movements using the other joints of the lower limb. Sometimes these joints are weak or damaged and a massage using this massager may be unpleasant or cause additional pain to such a patient. The device described herein could be improved by using an actuator to make the rollers move automatically. Additionally, the rollers could be arranged in planes parallel to each other and alternately apply pressure to different portions of the foot.

As mentioned above, the use of robots is possible in the rehabilitation of patients with musculoskeletal and nervous system disorders. Mechanical properties of such and similar devices (surgical robots, prostheses) are initially evaluated by performing simulations using the finite element method. In the program in which the simulations are performed, a tetrahedral finite element mesh is generated and the values of stresses, displacements, and strains are then determined. In addition, it is possible to perform simulations in the field of dynamics, such as natural frequency studies. In the paper [8] the authors have examined the natural frequency and mode shapes of the arm and the working tip of the da Vinci robot. An increase in vibration amplitude can cause the tool to move on the wrong track. This is detrimental to the health of the patient and the operation of the device. Based on the conducted tests it was determined for which natural frequencies the values of these amplitudes increase.

The model illustrating the principle of operation of the above-mentioned device for neuropathy's treatment (see Fig. 2) was created with the use of the Solid Works program. Finite element simulations were then performed using a linear elasticity theory to determine the stresses and displacements that would occur under the given loads occurring with the use of the device.

2 Neuropathy

Neuropathy is defined as the presence of subjective or clinical features of peripheral nervous system damage occurring in the course of various illnesses. The most common neuropathy occurs as a consequence of diabetes, but this ailment is also observed for example as a complication after chemotherapy. Analyzing the symptoms and complaints reported by patients, it can be noted that the disease hurts the nervous system, both in sensory, motor, and autonomic aspects. Sensory symptoms include pain, paresthesia, and a feeling of numbness, night falls, unsteady gait, while motor symptoms include muscle weakness, muscle atrophy, balance disorders, and ataxic gait. Initially, the nerve fibers with the longest axons, located in the lower extremities, are damaged. The patient perceives a sensory disturbance that is referred to as the "sock effect". When the neuropathy spreads, sensory disturbances are also observed in the upper extremities [3].

The reasons for neuropathy are diversified and depend on the condition that causes the ailment. For example, diabetic neuropathy is the result of a deficiency in metabolic products necessary for nerve cell function, and the neuropathy after chemotherapy results from damage to nerve fiber axons by cytostatics [3, 7].

There are many pharmacological approaches to treat or prevent neuropathy. Some medications, however, adversely affect the treatment process for the patient's chronic disease of which neuropathy is a consequence. In addition to pharmacological methods, patients may be treated with kinesitherapy. It is also safe for cancer patients. However, this therapy requires constant supervision by a physiotherapist in most cases due to the possibility of the patient losing balance due to sensory disturbances. Rehabilitation of people with peripheral nerve injuries is a process that takes time. The patient must also demonstrate a high level of commitment. Thanks to the therapy, the regeneration process of the nerve may be accelerated. It is also very important to prevent muscle atrophy and recurrent spasms. Appropriately started rehabilitation allows to maintain the full range of motion as much as possible and to avoid the fixation of abnormal movement patterns. For patients who do not have oncological disease, heat therapy treatments such as muscle electrostimulation, lignocaine iontophoresis can be used. This helps regenerate damaged nerve fibers. Additionally, cathodic and anodic galvanization, short-wave diathermy, or infrared radiation are used to stimulate fiber growth. For cancer patients, only manual therapy is used, such as massage, active and passive exercises, and lymphatic drainage [7].

3 Construction of the Novel Medical Device

The designed device has a massager function. In Fig. 2, an illustrative model of a device for rehabilitation of patients suffering from neuropathy is presented. Its main mechanism consists of five components. One component consists of a roll, a driveshaft, 2 bearings, and 2 supports. A geared motor is located at the rear of the device to provide the drive source. The proposed motor is a 25D × 48L HP 4. 4:1 6 V 2220 RPM model - Pololu 1570. The power transmission is realized by using spur gears.

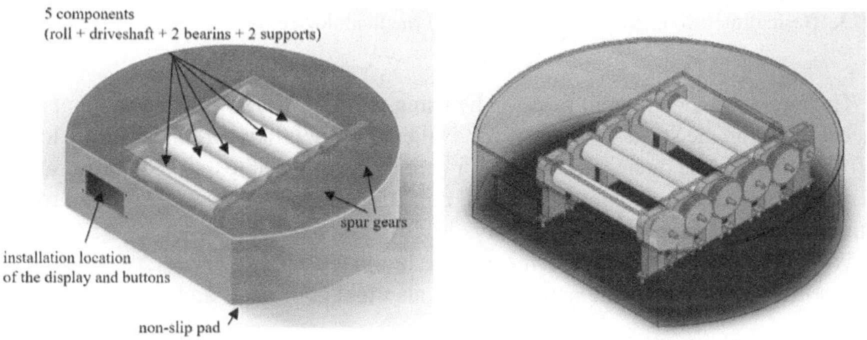

Fig. 2. The illustrative model of a novel medical device for neuropathy's rehabilitation.

The idea of the mechanism used resembles the principle of operation of an eccentric. The value of eccentricity was not consulted with health care professionals and was chosen proportionally to the dimensions of the construct in the test model. However, it may in the future depend on the patient's impressions of the pressure of the roller against his foot. The rolls are mounted eccentrically on the driveshaft. This allows not only to rotate the roll but also to move the selected point on its perimeter up and down. It is very important to ensure the correct initial position of the roll concerning the driveshaft. The first, third, and fifth driveshaft should be situated above the roll's axis. In the case of the other driveshafts, the situation is the opposite. After the device has been started, this construction can create a motion effect that resembles a "wave flow".

The basic dimensions of the device are presented in Fig. 3. The base of the device shall be provided with a non-slip pad. Due to the need for direct contact between the patient's foot and the roller surface, sterilization will be required. In order not to expose the rolls to wear and tear and the need for later replacement, it is planned to use a silicone layer, located in the hole in the upper part of the housing.

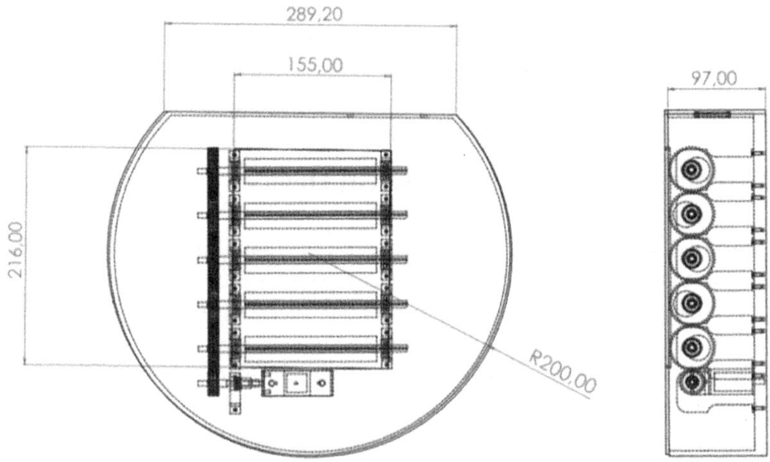

Fig. 3. Basic dimensions (unit: mm) of the novel medical device for neuropathy's rehabilitation.

The recommended patient position by using the device is sitting. The patient should place the foot on the surface of the silicone layer at the height where the rolls are located. Before activating the device, the patient or caregiver will be able to select the duration of the rehabilitation. The possibilities are 10 min, 15 min, or 20 min. The speed of rotation of the rollers will be a constant quantity.

With therapy from this device, the patient should feel less numbness, their feet should have more blood supply. The patient should also feel pain relief and relaxation.

4 Simulation of Mechanical Properties

4.1 Boundary Conditions and Materials Parameters

It is assumed that the patient will use the device for its intended purpose, following the instructions and recommendations of the medical professional. The moment when pressure is exerted by the patient's foot on all rolls was chosen. As a simplification, it was assumed that the load distribution would be uniform across all rolls. The patient's weight was assumed to be 150 [kg] and the selected load value per 5 components was the load corresponding to half the body weight. As calculated, this load has a value of 735,75 [N].

The study involved both the entire device and a single component consisting of a roll, a shaft, two bearings, a gear, and two supports. The first simulation was performed for the entire device (5 components).

Table 1 contains the materials proposed for the selected part of the device and Table 2 presents the properties of those materials. Properties of materials were taken from the library of the Solid Works program. The value of yield strength of ABS was taken from the external source.

Table 1. The materials for the selected part of the device.

Part's name	Material
Housing	Aluminum alloy 1060
Roll	ABS
Driveshaft	Chromium-plated stainless steel
Bearing	Steel 1.2311 (40CrMnMo7)
Support	Aluminum alloy 3003-H14

Table 2. Properties of materials used in the device model.

	Unit	Aluminum alloy 1060	ABS	Chromium-plated stainless steel	Steel 1.2311	Aluminum alloy 3003-H14
Density	[kg/m^3]	2700	1020	7800	7800	2730
Young's modulus	[N/mm^2]	69000	2000	200000	205000	69000
Poisson's ratio	[–]	0,33	0,394	0,28	0,28	0,33
Yield strength	[N/mm^2]	27,5742	30	172,339	821	145
Ultimate tensile strength	[N/mm^2]	68,9356	30	413,613	992	150

In the first simulation, the device's base was chosen as the constraint (Fig. 4). The load was applied to the outer surfaces of the rolls (perpendicular to the roll's axis), which is in contact with the patient's foot through a silicone layer. In the second simulation, the load's distribution is similar to the previous simulation. As the constraint, there were chosen lower surfaces of the supports.

4.2 Governing Equations of Linear Elasticity

Hooke's law for linear elasticity [9] is usually written like:

$$\mathbf{S} = \mathbf{C} : \boldsymbol{\varepsilon}, \tag{1}$$

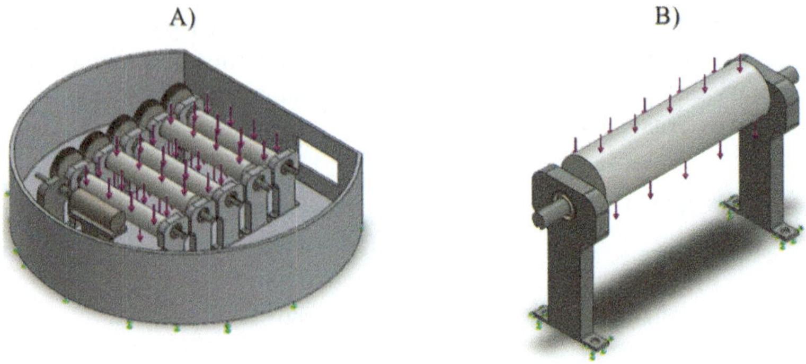

Fig. 4. The distribution of the loads and constraints in: A) the entire device, B) the single component.

where the stress tensor **S** and the strssain tensor **ε** are second-order tensors, while the constitutive tensor **C** is a fourth-order tensor. In a notation where the indices are shown, the same equation would read

$$S_{ij} = C_{ijkl} : \varepsilon_{kl}, \tag{2}$$

where the Einstein summation convention has been used.

In the case of the isotropic material the elasticity matrix becomes

$$\mathbf{C} = \frac{E}{(1+\nu)(1-2\nu)} \begin{bmatrix} 1-\nu & \nu & \nu & 0 & 0 & 0 \\ \nu & 1-\nu & \nu & 0 & 0 & 0 \\ \nu & \nu & 1-\nu & 0 & 0 & 0 \\ 0 & 0 & 0 & \frac{1-2\nu}{2} & 0 & 0 \\ 0 & 0 & 0 & 0 & \frac{1-2\nu}{2} & 0 \\ 0 & 0 & 0 & 0 & 0 & \frac{1-2\nu}{2} \end{bmatrix} \tag{3}$$

The equilibrium equation expressed in stresses is

$$-\nabla \cdot \mathbf{S} = \mathbf{F}, \tag{4}$$

where **F** denotes the volume forces (body forces). The equilibrium equation with zero body forces expressed in stresses for 3D (six equations and six unknowns S_{ij}) is

$$-\nabla \cdot \mathbf{S} = \mathbf{0}. \tag{5}$$

Substituting the stress-strain and strain-displacement relationship for linear elastic material in the above equation results in Navier's equation expressed in displacements formulation (three equations and three unknowns u_i)

$$-\left(\mu \nabla^2 \mathbf{u} + (\lambda + \mu) \nabla \nabla \cdot \mathbf{u}\right) = \mathbf{0}, \tag{6}$$

where λ and μ are Lamé parameters and can be expressed in terms of two elastic constants a $\lambda = \frac{E\nu}{(1+\nu)(1-2\nu)}$ and $\mu = \frac{E}{2(1+\nu)}$, where E is Young's modulus and ν is the Poisson's ratio.

4.3 Numerical Results

Both of the aforementioned simulations (Fig. 4) were initially performed using a finite element method [10] with an automatically generated grid (Fig. 5).

There were created the finest grids, as was possible. Due to the thin walls of the device, only a single component was analyzed with coarse mesh.

Table 3 contains the parameters of grids by type of simulation (Fig. 5). FEM was used to solve the linear elastic problem described by Navier's equation.

Table 3. The parameters of automatically generated grids (fine-grid meshes).

Simulation	Nodes [−]	Elements [−]
Entire device	148963	81614
Single component	87959	55773

The first simulation is more illustrative. It allows observing in which elements of the structure the highest stresses and displacements can be expected to be concentrated. The values of obtained stresses are in this case lower or close to the yield strength values. It is worth noting that the structure is most vulnerable in the middle part of the roll and the shaft, as well as in the places where the shaft is connected to the supports.

In the second, more accurate simulation, the highest stress value does not exceed the accepted yield stress values. It can therefore be concluded that none of the parts is particularly vulnerable to fracture damage. The displacement values obtained are relatively small and do not exceed the geometric dimensions of the part. Plots showing the stresses and displacements in the various elements of the device are presented in Figs. 6, 7, 8, and 9.

Table 4 contains the numerical results (maximum von Mises stress and maximum displacement) of two performed simulations.

Table 4. The numerical results of the simulations.

	Entire device	Single component
Load [N]	735,75	147,15
Max. stress von Mises [MPa]	30,46	16,51
Max. displacement [mm]	$3{,}859 \cdot 10^{-2}$	$1{,}41 \cdot 10^{-2}$

A)

B)

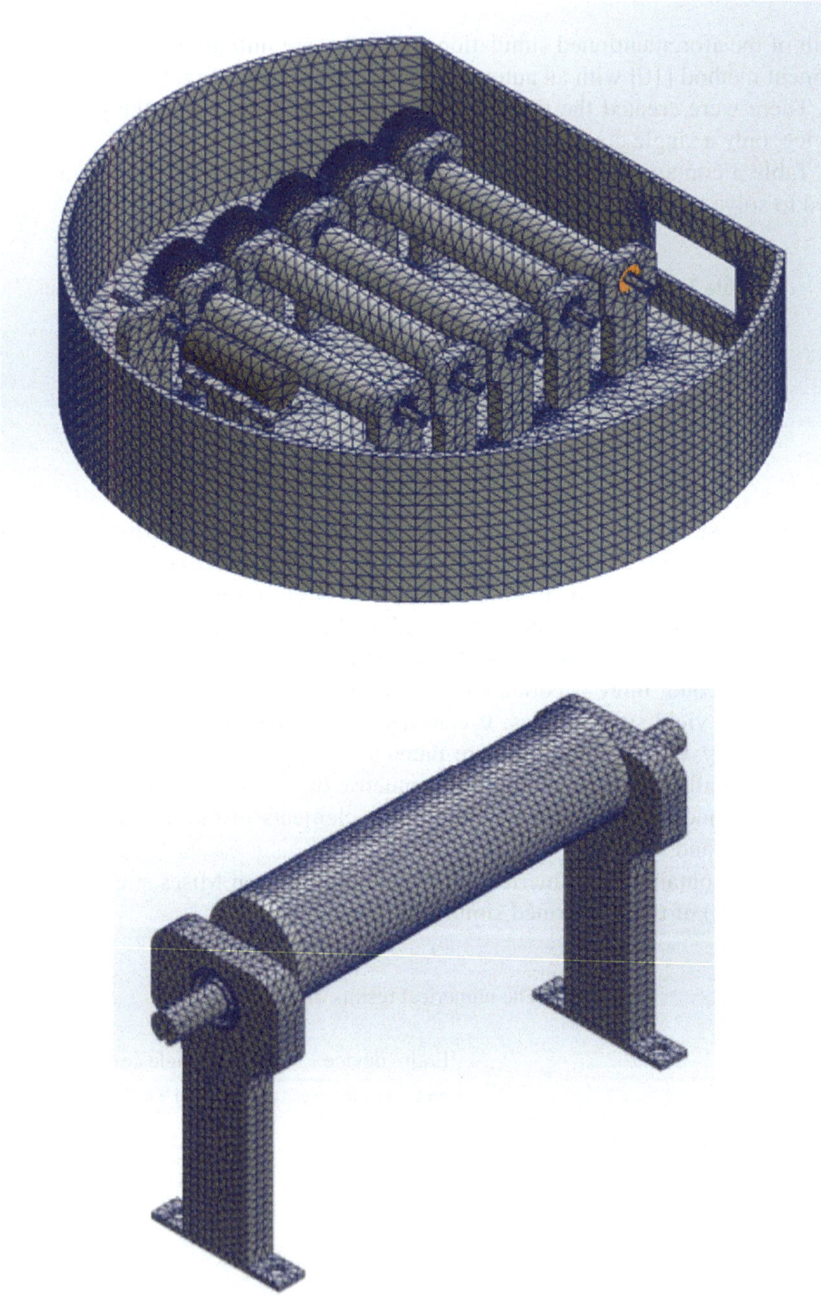

Fig. 5. The automatically generated grids (fine-grained meshes): A) the entire device, B) the single component.

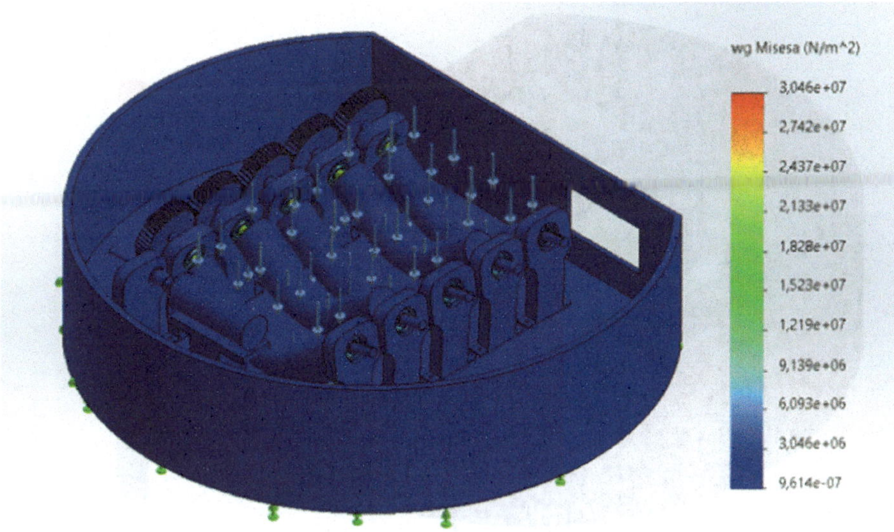

Fig. 6. Von Mises stresses [N/m^2] in the entire device.

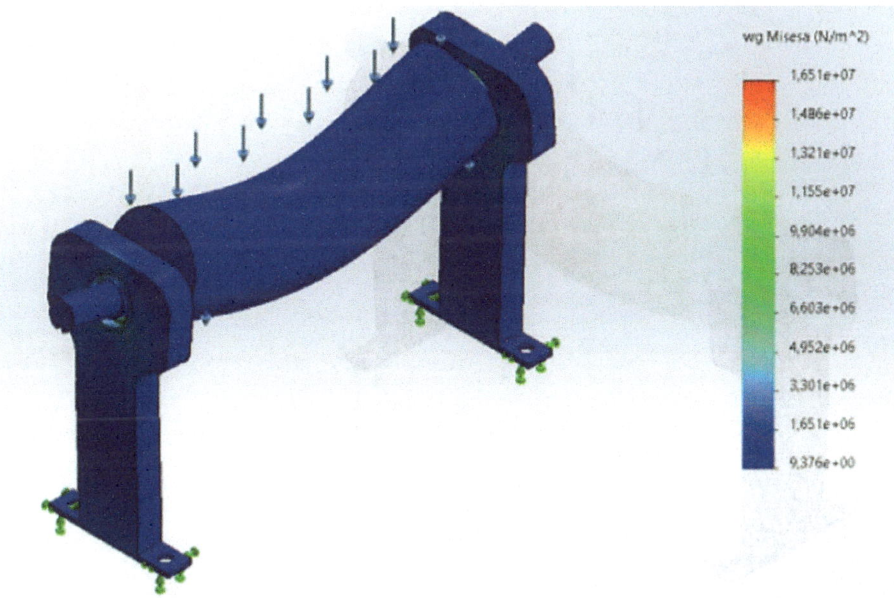

Fig. 7. Von Mises stresses [N/m^2] in the single component.

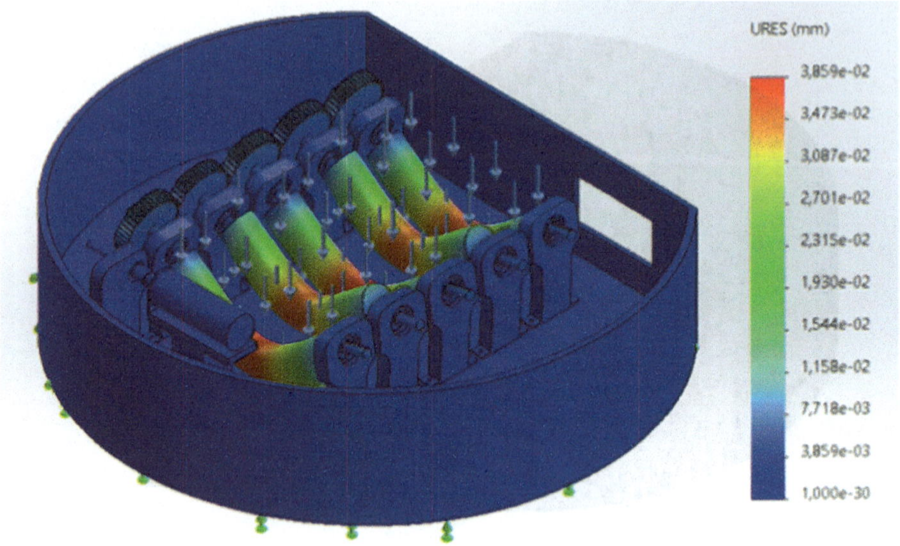

Fig. 8. Displacements [mm] in the entire device.

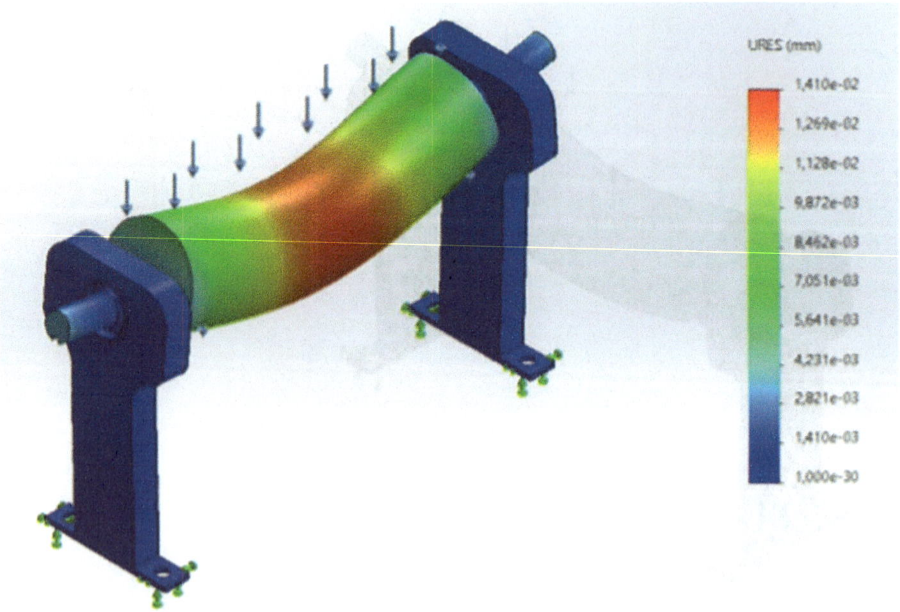

Fig. 9. Displacements [mm] in the single component.

5 Conclusions

Neuropathy is a consequence of many diseases and injuries. A sense of independence increases each patient's psychological comfort in the treatment process. Thanks to the designed device, the patient and his family will gain the possibility of home rehabilitation, without the need for frequent visits to the specialist. Taking into account the results obtained in each simulation, it can be concluded that the device can function safely and efficiently under the influence of the applied loads. For finite element simulations, the rolls were designed to have a smooth surface. Overly complex shapes could prevent the mesh from being generated correctly. The greater the surface irregularity would be, the more effective the massage of the neuropathy patient's foot would be.

Acknowledgments. This work was supported by grants of the Ministry of Education and Science in Poland: 0612/SBAD/3576 (2021/2022). The simulations have been carried out at the Institute of Applied Mechanics, Poznan University of Technology.

References

1. Burns, T.M., Mauermann, M.L.: The evaluation of polyneuropathies, Neurology **76**(7 Supplement 2), S6–S13 (2011)
2. Torpy, J.M., Kincaid, J.L., Glass, R.M.: Patient page: peripheral neuropathy. J. Am. Med. Assoc. **303**(15), 1556 (2010)
3. Szczyrba, S., Kozera, G., Bieniaszewski, L., Nyka, W.M.: Diabetic neuropathy - pathogenesis, diagnosis, prevention, treatment (in polish: Neuropatia cukrzycowa – patogeneza, rozpoznawanie, zapobieganie, leczenie), Klinika Neurologii Dorosłych Gdańskiego Uniwersytetu Medycznego, Gdańsk (2010)
4. Mavroidis, C., Nikitczuk, J., Weinberg, B., Danaher, G., Jensen, K., Pelletier, P.: Smart portable rehabilitation devices, Department of Mechanical & Industrial Engineering Northeastern University, Boston (2005)
5. Zhu, G., Zeng, X., Zhang, M., et al.: Robot-assisted ankle rehabilitation for the treatment of drop foot: a case study. In: Mechatronic and Embedded Systems and Applications (MESA), IEEE, pp. 1–5 (2016)
6. Pieber, K., Herceg, M., Paternostro-Sluga, T.: Electrotherapy for the treatment of painful diabetic peripheral neuropathy: a review. J. Rehabil. Med. **42**, 289–295 (2010)
7. Brzeziński, K.: Chemotherapy-induced peripheral neuropathy. Part II. Prevention (in polish: Obwodowa neuropatia wywołana chemioterapią. Część II. Zapobieganie), Współczesna Onkologia **16**(3), 262–265 (2012)
8. Geda, H., Kuliberda, A., Stręk, T.: Finite element analysis of natural frequencies and mode shapes of the da Vinci medical robot arm. Vibrat. Phys. Syst. **31**(2), 2020205 (2020)
9. Bower, A.F.: Applied Mechanics of Solids, CRC Press, Taylor & Francis Group, Boca Raton (2010)
10. Zienkiewicz, O.C., Taylor, R.L.: The Finite Element Method. Butterworth-Heinemann, Oxford (2000)

Product-Service System for the Pharmaceutical Industry - A New Opportunity for Machine Manufacturers

Mariusz Salwin[1(✉)], Andrzej Kraslawski[2], Michał Andrzejewski[2,3], and Jan Lipiak[4]

[1] Faculty of Mechanical and Industrial Engineering, Warsaw University of Technology, 85 Narbutta Street, 02-524 Warsaw, Poland
mariusz.salwin@onet.pl
[2] School of Engineering Science, Lappeenranta University of Technology, P.O. Box 20, 53581 Lappeenranta, Finland
[3] Kalmar-Pac, 38 Oksywska Street, 01-694 Warsaw, Poland
[4] Etigraf Printing House, 52 Głowackiego Street, 05-071 Sulejówek, Poland

Abstract. Nowadays, special attention is paid to developing new solutions and applying them to sectors of the economy that can ensure the fastest development and security. Product-Service System (PSS) is commonly understood as an innovative solution distinguishing enterprises from the competition and supporting environmental protection. Designing new PSS for new industries is a very important area of PSS research and development. Industrial practice and literature analysis do not provide examples of the use of PSS in the pharmaceutical industry. The aim of the article is to design a conceptual PSS for the pharmaceutical sector. The example presented in the article was developed at an industrial workshop that took place during the sale of a pharmaceutical machine. When creating PSS, the authors relied, on the one hand, on the capabilities of the company selling machines and providing their service, and on the other hand, on the wishes, needs and problems of the user of these machines. The developed PSS provides the pharmaceutical company with a wide range of services in response to its requirements. The research carried out in the form of workshops draw attention to issues important for the pharmaceutical industry and provide guidance for manufacturers and suppliers of machinery and equipment used in this area.

Keywords: Product-Service System (PSS) · Pharmaceutical industry · Pharmaceutical machine

1 Introduction

In the last thirty years, production enterprises have been rebuilding their offers in order to gain a competitive advantage [1, 2]. Producers go on a servitization journey, moving away from selling the products themselves to providing customers with an integrated mixture of products and services [3, 4]. An important component of servitization is Product-Service Systems (PSS) [5]. It is an innovative solution that constitutes a new market

F. Gorski et al. (Eds.): MANUFACTURING 2022, LNME, pp. 120–135, 2022.
https://doi.org/10.1007/978-3-030-99769-4_10

proposition, which, thanks to the inclusion of additional services in the offer, extends the functionality of the product [6, 7]. PSS-based solutions significantly increase the value for the user and are more attractive to them. One of the essential features of PSS is the unique adaptation of solutions to the ever-changing and more sophisticated needs of the client. By using PSS, traditional manufacturing companies distinguish themselves from their competitors [8, 9].

PSS design activities face various challenges. One of such challenges is the high personalization of solutions, which is a response to specific wishes, needs and problems of the client [10, 11]. This means that customers expect to meet their expectations and help build their own PSS vision. In developing personalized solutions, the manufacturer is to be a helpful partner that accurately records and analyzes all customer requirements [12–14]. Meeting customer requirements will allow for the development of a successful PSS. This guarantees higher revenues for the producer and enables building lasting relationships with customers. A well-designed PSS reduces the consumption of materials and electricity, which makes it environmentally friendly [15, 16]. The ability of enterprises to design and implement successful PSS is recognized as a key success factor. Therefore, an important area of research on PSS is design [17, 18].

The pharmaceutical sector is the characteristic branch of the economy [19, 20]. It is responsible for the drug safety of individual countries. It includes the development and production of medical products, devices, machines and format parts as well as raw materials and materials used in production. It is governed by a number of legal regulations and safety standards to which all parts of the supply chain and production processes are subject [21–23]. The sector is characterized by innovation and high research expenditure. In the time of the COVID-19 pandemic, the pharmaceutical sector is of key importance to the economy. The research and development of a vaccine or cure for coronavirus is a test for the industry. The crisis triggered by the pandemic shows that the world needs a strong pharmaceutical sector that needs to be supported.

Scientists, drug manufacturers, suppliers of raw materials, laboratory equipment, machines, packaging and solutions used in the pharmaceutical industry gathered at the annual Congress of the World of the Pharmaceutical Industry addressed topics that will have a major impact on the industry in the coming years. The Analyzes of the needs and problems of producers regarding the safety and quality of drugs, modern technologies, work and production optimization have shown that the pharmaceutical sector needs comprehensive solutions in this area. The answer to the topics discussed by the participants of the congress may be PSS for pharmacy. This solution may turn out to be a path that will accelerate the development of this sector [24, 25].

The aim of this work is to develop a conceptual Product-Service System model for the pharmaceutical sector. The research was carried out in the form of research workshops during the sale of a pharmaceutical machine. The workshop was attended by a machine sales and service company and a pharmaceutical company.

The paper is structured as follows: the first part is the introduction. The next part contains the research methodology. The third part presents the literature analysis. The next part presents results obtained at research workshops. The last part is the discussion and conclusions.

2 Research Methodology

2.1 Research Aim

The purpose of this article is to develop a Product-Service System for the pharmaceutical industry. The following research questions were asked in the paper:

- What are the possibilities of using the PSS in pharmacy?
- Can the use of the PSS bring benefits to manufacturers of pharmaceutical machinery and equipment and companies producing medicinal products?

The research carried out in the article highlights a set of guidelines that can be used in industrial practice for the development of the PSS in the aforementioned sector of the economy.

The research methodology adopted in this article consisted of four main phases:

1. Systematic literature review - focuses on two main phases performed in parallel. The first is a systematic literature review of the PSS industrial cases, the second is a systematic literature review of PSS design methods. The years 2001–2019 were assumed as the period of analysis. The sources of information selected for these studies were scientific databases (including Web of Science, Scopus, Science Direct, Springler Link). This database includes major and minor publishers. The keywords selected for the search are "Product-Service System in industry" or its synonyms. The analysis included scientific articles from journals, book chapters, conference articles, industry reports and white papers in English. The result was finding 150 works describing PSS functioning in the industry. Assuming the same analysis period and using the same databases, the authors conducted a systematic review of the literature on PSS design methods. The keywords selected for the search are Product-Service System design "or its synonyms. The result was a finding of 60 methods in 64 works.
2. Pharmaceutical industry research. In this phase, the manufacturers of pharmaceutical machines and their users were analyzed. This phase also includes the analysis of the pharmaceutical industry reports.
3. Research and analysis of enterprises. The research was carried out in the form of research workshops during the sale of a pharmaceutical machine. In this phase, the needs and problems related to the pharmaceutical machine used in the production of tablets and the demand for services in a pharmaceutical company were analyzed. Pareto analysis, brainstorming and questionnaire were used in the workshops. Pareto analysis allowed to identify the causes causing costs and financial losses for the company. The brainstorming allowed for the identification of the company's needs related to services. Based on the information obtained, a questionnaire was developed. The survey was conducted among the company's employees to find services corresponding to the needs and problems of the company.
4. Construction of the PSS for the pharmaceutical industry. Based on the conducted research, the PSS for pharmaceutical machines was designed. The developed PSS includes services that meet the needs of the enterprise and help to solve its problems. This is the best solution for the analyzed enterprise.

3 Literature Review

3.1 Product-Service System in Industrial Practice

This stage of the analysis concerned the PSS models used in industry. The examined PSS are developed by international corporations and are addressed to various sectors of the economy. The hallmarks of these models include advanced product technology and high product value. They are very often machines composed of many systems divided into main parts and auxiliary parts. This means that in order to ensure continuous use of these products, professional service is necessary. The long life cycle of the products is also an important element. The main examples of the PSS are Rolls-Royce (power-by-the-hour for engines) and Azimut Yachts (Maintenance Service Program). The conducted analysis of the PSS used in industry did not reveal PSS used in pharmacy.

3.2 Product-Service System Design

Designing the PSS is one of the key areas of the PSS research. Literature analysis by systematic literature search identifies 60 different approaches to PSS design. From the 60 analyzed methods, 12 have been successfully verified in industrial practice, and 21 verified in scientific projects. Each of the analyzed methods presents the industries in which the case studies were performed. Therefore, a list of PSS design methods and the industries in which they are used was developed (Fig. 1).

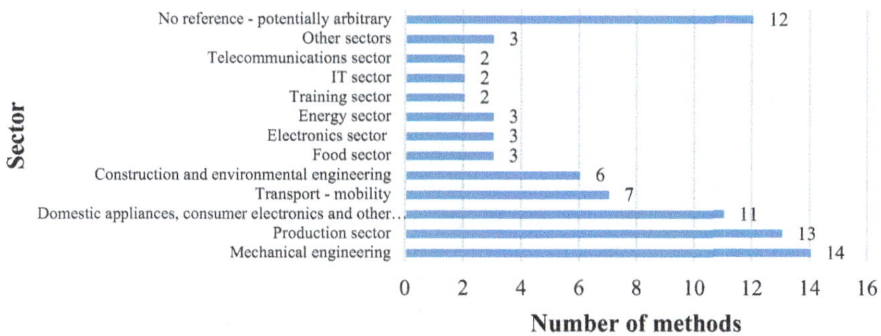

Fig. 1. Classification of PSS design methods by sector [26].

Out of all 60 analyzed methods, as many as 14 are used in the branch of economy, which is mechanical engineering. It is impossible to assign 12 methods to any of the sectors of the economy. These are universal approaches to the PSS design that can be implemented in any industry. Several methods are used in quite different sectors of the economy. The available literature does not provide guidelines or a framework for the design of the PSS for the pharmaceutical industry [26].

3.3 Pharmaceutical Industry

Pharmacy is responsible for the research, development and manufacture of medicinal products. It is one of the most innovative sectors of the economy that develops new drugs based on long and complex scientific research. It is crucial for the timely delivery of drugs that reach pharmacies and hospitals. Products from this industry are characterized by high quality and the use of active substances. They are used to save the health and life of people and animals. The production of medicinal products is subject to many restrictions related to the hygiene of production and the protection of intellectual property.

The largest pharmaceutical markets are North America, Europe and Asia. In 2018, the revenues of the 10 largest pharmaceutical markets in the world amounted to EUR 796 billion. In the European Union, 595,751 people work in 4106 companies in the pharmacy sector. The value of production in the European Union alone in 2018 amounted to EUR 287.89 billion (Fig. 2).

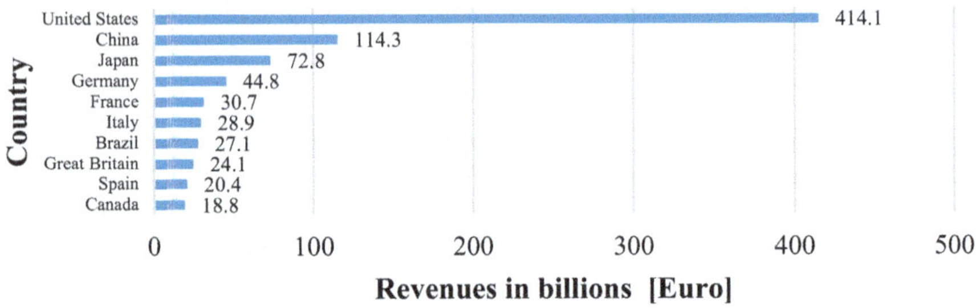

Fig. 2. Revenue from the top 10 domestic pharmaceutical markets in the world in 2018 and 2019 [27].

3.4 Pharmaceutical Machines

Pharmaceutical machines are very important element of the pharmaceutical sector. They are elementary capital goods, because the medicinal products produced on them are needed to stop the epidemic and save people's lives and health.

Machines and devices used in pharmaceutical production (including tablet presses, capsule presses, blister machines, pharmaceutical powder mixers, dryers, bench presses) are the key element used in the production of drugs and a basic element of equipment for drug manufacturers. Pharmaceutical machines differ in the level of automation, size and method of operation. A specific drug form (e.g. a tablet) can only be produced on a specific machine (tablet press).

Manufacturers (for example Uhlmann, Norden, Citus Kalix, HAPA, PKB) offer machines and devices used in pharmacy in various versions together with basic services (transport, presetting, installation). The average value of investments of pharmaceutical companies operating in the European Union in machinery and equipment in 2014–2018 amounted to 5.91 billion EUR (Fig. 3).

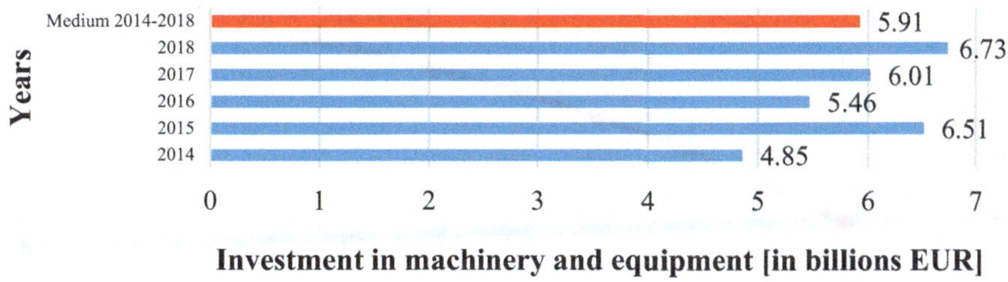

Fig. 3. Investments in machinery and equipment of pharmaceutical companies in Europe Union [in billions EUR] [28].

4 Results

4.1 Characteristics of the Analyzed Company

The analyzed enterprise is a producer of drugs and medical devices with experience and tradition. It mainly specializes in the production of generics and over-the-counter drugs. The plant has specialized laboratories and modern machinery. The company has qualified human resources with extensive experience. Our care for the high quality and effectiveness of our products means that they have been trusted by consumers, patients and pharmacists for years. Among the range of products manufactured by the analyzed company, the following can be distinguished: antibiotics, drugs for influenza, digestion, circulation, painkillers and non-invasive medical devices. The company is constantly developing and conducting research into the development of new drugs and their production.

4.2 Company Problems and Needs

The aim of this stage of the research was to identify the needs and problems faced by a pharmaceutical company, i.e. a user of pharmaceutical machines used in production. The analysis took place in the form of a research workshop. They were combined with a meeting at which the analyzed plant purchased the machine. A representative of a company dealing in the sale and service of machines used in the pharmaceutical industry also participated in the study.

4.2.1 Company Problems

Despite the tradition and experience, the company struggles with a number of problems related to the production of its products. In order to diagnose them, the Pareto-Lorenz analysis was used (Fig. 4). The improvements implemented by the company all the time reduce the problems. Nevertheless, they generate high significant costs. In 2019, the most important ones included: retooling of pharmaceutical machines, rejects during production, test runs and failures of pharmaceutical machines, non-compliance of the finished product with its specification, service of pharmaceutical machines, training and production contamination.

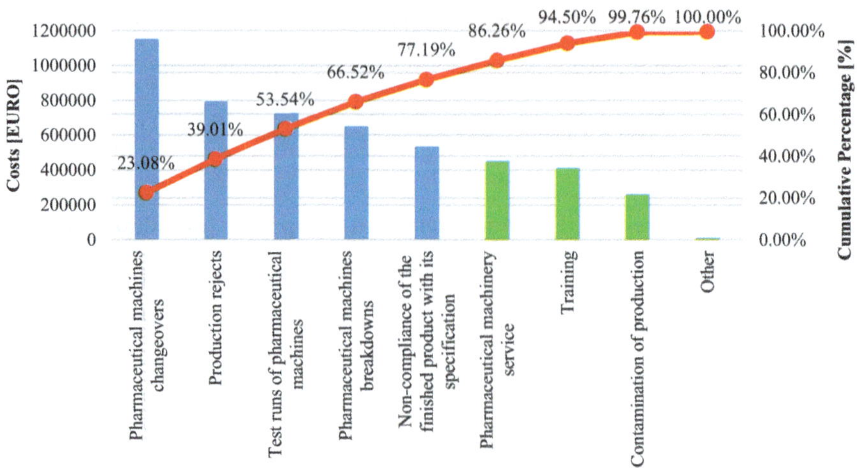

Fig. 4. Pareto-Lorenz analysis of losses in the analyzed pharmaceutical company.

The problems that bring the greatest losses are pharmaceutical machines changeovers, production rejects, test runs of pharmaceutical machines, pharmaceutical machines breakdowns, non-compliance of the finished product with its specification. This is as much as 77.19% of all losses. These factors negatively affect the availability of machines and the quality of production. As a result, they reduce the Overall Equipment Effectiveness, which during the working day is about 65%, while in 2019 it was on average about 59%.

All losses presented here are specific to the pharmaceutical sector. Production in this sector is characterized by very high precision and cleanliness. In both cases, even the smallest deviations from the standards prevent the product from leaving the production hall and reaching the customer. Additionally, it generates rejects during production. Operators are important figures in the entire production process. They are responsible for the operation of machines and a number of related parameters.

4.2.2 Company Needs

The conducted workshops allowed for the diagnosis of the main needs of a pharmaceutical company, concerning:

- machine purchase - elimination of high purchase costs, machine rental with operation and service;
- use and service of a pharmaceutical machine - service subscription, replacement of spare parts, proper inspection of format parts, cleaning and washing, retooling, error diagnostics;
- training - the purchase of a new machine requires re-training of the operating operators. In addition, various types of training, including the use of machines, must be subjected to every new person employed in the production department;

- drug production process - compliance with sanitary and hygienic requirements, proper handling of the product, elimination of typically mechanical problems related to the need to shape the products properly, cleaning, changeovers, fault diagnosis, production support - trained support staff;
- sanitary and hygienic requirements - guaranteeing specific hygiene conditions that meet the highest stringent quality requirements, machine washing procedures, division into clean and dirty zones, adequate protection of employees;

The presented areas are very important for the analyzed enterprise. They affect its production capabilities and production quality. In some cases, they generate financial and raw material losses.

4.3 The Service Needs of the Pharmaceutical Company

The next phase of the workshops concerned the selection of services that the analyzed enterprise needs. These services will be part of the PSS for pharmaceutical machines. All this so that the new PSS responds to the needs and requirements of the analyzed company, and also helps to solve emerging problems or completely eliminate them. Four groups of services have been selected:

- universal related to pharmaceutical machine;
- related to the production of drugs;
- related to health and safety at work;
- additional;

Representatives of the analyzed enterprise from each group of services chose those which were the most important from their point of view (Table 1).

Table 1. Services selected by the pharmaceutical company's employees from the broadened areas.

	Services related to a pharmaceutical machine	Services related to a drug manufacturing	Services related to health and safety at work	Additional services
Management board	Financial services	Optimization of changeovers of format parts	Training	Take-bake
	Rent	Optimizing the supply of raw materials and materials	Advice, consultations and training in pharmaceutical law	Optimizing the use of utilities (water, air, electricity)
	Training	Quality control of finished products	Safety certificates and sets of standards used in pharmaceutical production	Audits

(continued)

Table 1. (*continued*)

	Services related to a pharmaceutical machine	Services related to a drug manufacturing	Services related to health and safety at work	Additional services
	Delivery, installation, commissioning	Giving the right shape to products	Sterilization	
	Warranty	Waste disposal	Waste disposal	
	Machine software update			
	Updating (reconstruction, modernization) of the machine			
Middle management	Training	Optimization of changeovers of format parts	Training	Lean tools
	Warranty	Giving the right shape to products	Advice, consultations and training in pharmaceutical law	Disposal of protective materials
	Supply of spare parts	Packing and cost optimization of packaging materials	Security checks	Take-bake
	Maintenance and inspection	OEE analysis and optimization	Optimization of the supply of protective materials	Audits
	Service agreement		Optimizing job matching	
	Updating (reconstruction, modernization) of the machine		Optimization of work ergonomics	
	Diagnostics and troubleshooting		Sterilization	
	Monitoring, machine operation			
Production staff	Training	Optimization and standardization of working time	Training	Disposal of protective materials
	Washing and cleaning	Training and integration	Optimizing job matching	Optimization and standardization of working time
	Repair shop equipment	Data visualization on cards and boards	Optimization of work ergonomics	Integration of data visualization on cards and boards

(*continued*)

Table 1. (*continued*)

	Services related to a pharmaceutical machine	Services related to a drug manufacturing	Services related to health and safety at work	Additional services
	Supply of spare parts	Cleaning and washing format parts	Noise reduction	
	Maintenance and inspection		Safety certificates and sets of standards used in pharmaceutical production	
	Diagnostics and troubleshooting			

During the research, the analyzed company confirmed its interest in renting a machine with services under the service subscription.

4.4 Product-Service System for Pharmaceutical Industry

This chapter deals with model development. At this stage, the PSS for pharmaceutical machines are built. It was developed on the basis of the research carried out in the company, the suggestions of the company dealing with the sale and service of machines and the analysis of the literature on the PSS design. In the system under development, the main parties will be the pharmaceutical company and the machinery manufacturer or his representative. The main components of the PSS include the machine used in pharmaceutical production and services that meet the requirements of the analyzed enterprise (Table 2).

Table 2. Service packages in Product-Service System for pharmaceutical industry.

Area	Elementary set of services	Intermediate elementary set of services	Advance elementary set of services	Exclusive elementary set of services
Operation and service of the machine	Rent	Washing and cleaning	Supply and service of format parts used in production	Updating (rebuilding, upgrading) the machine
	Delivery, installation, commissioning	Delivery of spare parts	Diagnostics and debugging	Monitoring, operation of the machine

(*continued*)

Table 2. (*continued*)

Area	Elementary set of services	Intermediate elementary set of services	Advance elementary set of services	Exclusive elementary set of services
	Warranty	Maintenance and inspections	Take-bake	Update of machine software
Training	Trainings related to the operation of machines	Pharmaceutical health and safety training	Safety certificates and sets of standards used in pharmaceutical production	Advice, consultation and training in pharmaceutical law
Production process	Optimization of changeovers of format parts	Product shaping	Optimization and standardization of working time	OEE analysis and optimization
	Optimizing the supply of raw materials and materials	Packaging and cost optimization of packaging materials	Data visualization on cards and tables	Quality control of finished products
Occupational health and safety in pharmacy	Optimization of the supply of protective materials	Optimization of workstation layout	Optimization of work ergonomics	Sterilization of production departments
	Division of production zones into clean and dirty	Segregating risk factors according to their hierarchy	Noise reduction	Maintenance and improvement of health and safety management system
Environment	Segregation and utilization of waste and protective materials	Building a sustainable supply chain	Optimization of media use (water, air, electricity)	Designing products and processes in a way that limits their impact on the environment
Additionally	Financial services	Insurance services	Audits	Legal advice and consultation

A machine manufacturer supplies a pharmaceutical company with a specific machine. Together with the machine, the pharmaceutical company receives a package of services from which it can choose the services it needs most (Table 2 and Fig. 5). The manufacturer of the machinery charges a monthly fee consisting of three elements: a rental fee, a set monthly fee for the number of products produced and a service fee. Ownership rights remain with the machine manufacturer. The use of the machine, i.e. the production of as many of the pharmaceutical products concerned as possible, is the main need of the company in question. This shows that ownership is not a key need. The machine can be exchanged for a new one or upgraded or reconditioned after a mutually agreed period of use.

Table 3. Product-Service System for pharmaceutical industry—main assumptions.

Ownership	Sales	Services	Advantages to the manufacturer	Benefits for the customer
It remains with the machine manufacturer or its representative	Rent based on a monthly subscription	Elementary set of services	Possibility of reusing the machine in another company	Focus on new drug development
	Number of products manufactured	Intermediate elementary set of services	Improved customer relations	Increase in production capacity
Not transferred to a pharmaceutical company		Advance elementary set of services	Monitoring control and maintenance of the machine	Cost reduction
	Service subscription	Exclusive elementary set of services	Reduce resource consumption	Meeting stringent requirements for pharmaceutical production

The model developed has benefits (Table 3). The pharmaceutical company does not invest huge financial resources in the machine. In addition, it focuses mainly on its core business. Service, maintenance and repair issues are not within the scope of the user. The pharmaceutical company receives additional support in the form of services related to the production of drugs, which makes it possible to increase work efficiency and faster production. An additional benefit is the provision of various training courses, consultations and advisory services, which save the customer time and financial resources that it would otherwise have had to spend on its own. Occupational health and safety services support the machine user in meeting the stringent criteria of pharmaceutical production.

A pharmaceutical machine manufacturer makes money at the same time from the production of the machine, its use by the customer and related services. Precise manufacturing, high-quality materials and modern technological solutions are important points thanks to which the manufacturer can extend the life cycle of the machine. Additionally, thanks to this solution, the manufacturer monitors the machine, collects data and analyses its operation. All of this can translate into better action when it comes to upgrading or reconditioning the machine and contribute to a significant improvement in new machine series. It is important that components which, after a certain period of use, do not comply with the strict parameters and components which wear out quickly can be replaced efficiently. All this is done in order to meet the stringent requirements of the pharmaceutical industry and to ensure that the machine can operate continuously.

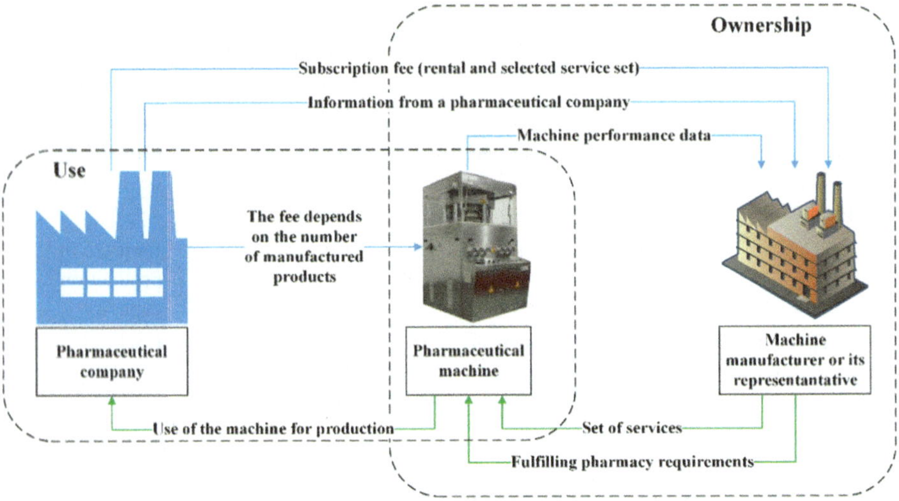

Fig. 5. Product-Service System for pharmaceutical industry—a concept.

5 Discussion and Conclusion

This paper analyzes the pharmaceutical industry. This market is characterized by high expenditure on research and development and implementation of innovations. Micro and small enterprises dominate here, but large enterprises invest the most in research and are responsible for inventing and patenting new medicinal products. The nature of this sector enables the use of completely new solutions related to the offer of manufacturers of machinery and equipment. It should be noted that the current offer of machine manufacturers operating in this industry in the field of the PSS is very narrow.

The Product-Service System is a solution consisting of products and accompanying services. The PSS aims to meet the requirements and needs of customers and to provide support in solving their problems. Currently, it is important to use the opportunities offered by the solution in business practice. The paper presents research workshops in which PSS was designed for pharmaceutical machines used in the production of drugs. The information obtained during the workshops made it possible to assemble a pharmaceutical machine and related services at the PSS.

The main findings of the workshops carried out include:

- knowing the wishes, needs and problems of the client and analyzing them is a flywheel for the PSS design team. Customer requirements are the guideline in the selection of services at PSS. It is important to involve the user in the design of the PSS
- an important element that should be taken into account when designing the PSS for this industry are: stringent requirements and legal regulations, pharmaceutical safety regulations including CGMP, FDA CFR 21 part 117 and occupational hygiene. It is also a challenge for the PSS in pharmacy.
- pharmaceutical use of the PSS can increase the speed of drug development. Then, significantly accelerate their production by increasing efficiency and eliminating many losses that occur during production. This will avoid the situation of reducing the supply of preparations that are of key importance in the elimination of threats to human life and health.
- appropriate selection of services for a specific pharmaceutical company and a specific type of production may lead to a reduction in production rejects, which will increase the number of manufactured products.
- the PSS fee eliminates the high investment for a pharmaceutical company, as well as costs such as depreciation, taxes and more associated with traditional purchasing.
- thanks to the fact that the manufacturer retains the ownership right, it is possible to constantly control the parameters of filling and forming products and immediately react when they change. This will reduce rejects during production. In addition, it will enable the diagnosis of the weakest parts that are in the machine. The developed model will allow the manufacturer to gain knowledge that can be used in new machines.
- this model can be targeted at enterprises of all sizes. It will allow micro, small and medium-sized enterprises to develop faster, and it will give large enterprises the opportunity to focus more attention on developing and patenting new products.

Research workshops in which two companies from the pharmaceutical sector have been involved are the foundation for research on the PSS in pharmacy. They indicate aspects and elements that need to be analyzed more broadly in future research.

References

1. Uchihira, N., Kyoya, Y., Kim, S.K., Maeda, K., Ozawa, M., Ishii, K.: Analysis and design methodology for recognizing opportunities and difficulties for product-based services. J. Inf. Process. **16**, 13–26 (2008)
2. Uchihira, N., Kyoya, Y., Kim, S.K., Maeda, K., Ozawa, M., Ishii, K.: Analysis and design methodology for recognizing opportunities and difficulties for product-based services. In: PICMET 2007 - 2007 Portland International Conference on Management of Engineering & Technology, Portland, OR, USA, pp. 2755–2762. IEEE (2007)
3. Salwin, M.: Design of Product-Service Systems in printing industry, Publishing House of the Lappeenranta-Lahti University of Technology LUT, Lappeenranta (2021). https://lutpub.lut.fi/bitstream/handle/10024/163257/Mariusz%20Salwin%20A4.pdf?sequence=1&isAllowed=y

4. Pirola, F., Boucher, X., Wiesner, S., Pezzotta, G.: Digital technologies in product-service systems: a literature review and a research agenda. Comput. Ind. **123**, 103301 (2020). https://doi.org/10.1016/j.compind.2020.103301
5. Martinez, V., Bastl, M., Kingston, J., Evans, S.: Challenges in transforming manufacturing organisations into product-service providers. J. Manuf. Technol. Manag. **21**, 449–469 (2010)
6. Gaiardelli, P., et al.: Product-service systems evolution in the era of Industry 4.0. Serv. Bus. **15**(1), 177–207 (2021). https://doi.org/10.1007/s11628-021-00438-9
7. Annarelli, A., Battistella, C., Nonino, F.: Product service system: a conceptual framework from a systematic review. J. Clean. Prod. **139**, 1011–1032 (2016)
8. Baines, T., Lightfoot, H.: Servitization in the aircraft industry: understanding advanced services and the implications of their delivery. In: Lay, G. (ed.) Servitization in Industry, pp. 45–54. Springer, Cham (2014). https://doi.org/10.1007/978-3-319-06935-7_3
9. Annarelli, A., Battistella, C., Nonino, F.: Competitive advantage implication of different product service system business models: consequences of 'not-replicable' capabilities. J. Clean. Prod. **247**, 119–121 (2020)
10. Salwin, M., Kraslawski, A., Lipiak, J., Gołębiewski, D., Andrzejewski, M.: Product-service system business model for printing houses. J. Clean. Prod. **274**, 122939 (2020)
11. Andriankaja, H., Boucher, X., Medini, K.: A method to design integrated product-service systems based on the extended functional analysis approach. CIRP J. Manuf. Sci. Technol. **21**, 120–139 (2018)
12. Li, A.Q., Kumar, M., Claes, B., Found, P.: The state-of-the-art of the theory on product-service systems. Int. J. Prod. Econ. **222**, 107491 (2020). https://doi.org/10.1016/j.ijpe.2019.09.012
13. Lim, C.-H., Kim, K.-J., Hong, Y.-S., Park, K.: PSS Board: a structured tool for product–service system process visualization. J. Clean. Prod. **37**, 42–53 (2012)
14. Salwin, M., Kraslawski, A.: State-of-the-art in product-service system classification. In: Ivanov, V., Trojanowska, J., Pavlenko, I., Zajac, J., Peraković, D. (eds.) DSMIE 2020. LNME, pp. 187–200. Springer, Cham (2020). https://doi.org/10.1007/978-3-030-50794-7_19
15. Vezzoli, C., Kohtala, C., Srinivasan, A., Diehl, J.C.: Product-Service System Design for Sustainability. Green Leaf, Sheffield (2014)
16. Emili, S., Ceschin, F., Harrison, D.: Product-service system applied to distributed renewable energy: a classification system, 15 archetypal models and a strategic design tool. Energy Sustain. Dev. **32**, 71–98 (2016)
17. Medini, K., Boucher, X.: Value chain configuration for PSS delivery – evidence from an innovative sector for sludge treatment. CIRP J. Manuf. Sci. Technol. **12**, 14–24 (2016)
18. Salwin, M., Andrzejewski, M., Kraslawski, A.: Information asymmetry in product-service system design - example of the pharmaceutical industry. Procedia Manuf. **55**, 282–289 (2021). https://doi.org/10.1016/j.promfg.2021.10.040
19. Salwin, M., Kraslawski, A., Andrzejewski, M., Lipiak, J.: Product-service system for the pharmaceutical industry. In: Camarinha-Matos, L.M., Boucher, X., Afsarmanesh, H. (eds.) PRO-VE 2021. IAICT, vol. 629, pp. 485–493. Springer, Cham (2021). https://doi.org/10.1007/978-3-030-85969-5_45
20. Valencia Cardona, A.M.: An Exploration of Smart Product-Service System Design: Guidelines and Insights for Design Management (2017). http://resolver.tudelft.nl/uuid:a78ba8b6-b865-4408-9788-468e635fa39b
21. European Federation of Pharmaceutical Industries and Associations: The Pharmaceutical Industry in Figures, Key Data 2015 (2015)
22. European Federation of Pharmaceutical Industries and Associations: The Pharmaceutical Industry in Figures, Key Data 2016 (2016)
23. European Federation of Pharmaceutical Industries and Associations: The Pharmaceutical Industry in Figures, Key Data 2017 (2017)

24. European Federation of Pharmaceutical Industries and Associations: The Pharmaceutical Industry in Figures, Key Data 2019 (2019)
25. European Federation of Pharmaceutical Industries and Associations: The Pharmaceutical Industry in Figures, Key Data 2020 (2020)
26. Salwin, M., Kraslawski, A., Lipiak, J.: State-of-the-art in product-service system design. In: Panuwatwanich, K., Ko, C.-H. (eds.) The 10th International Conference on Engineering, Project, and Production Management. LNME, pp. 645–658. Springer, Singapore (2020). https://doi.org/10.1007/978-981-15-1910-9_53
27. Top national pharmaceutical markets by revenue 2019. https://www.statista.com/statistics/266469/revenues-of-the-top-10-global-pharmaceutical-markets/. Accessed 27 Jan 2021
28. Eurostat - Data Explorer. https://appsso.eurostat.ec.europa.eu/nui/show.do?dataset=sbs_na_ind_r2&lang=en. Accessed 27 Jan 2021

Quality Estimation on the Application Process of the Vaccination Scheme Against COVID in Mexico

Rogelio J. Bautista-García[1(✉)], Rosa Ma Salinas-Hernández[2], Fidel Ulín-Montejo[2], and Manuel F. Suárez-Barraza[3]

[1] Smarthinking, Mexico City, Mexico
rogelio.bautista@smart-thinking.com.mx
[2] Universidad Juárez Autónoma de Tabasco, Villahermosa, Mexico
{rosa.salinas,fidel.ulin}@ujat.mx
[3] Universidad de las Américas Puebla, Puebla, Mexico
manuel.suarez@udlap.mx

Abstract. This work is based on the study of the quality of the application process of the vaccine via the skin against the SARS-COV-2 virus in Mexico, during the period from January to July 2021, through the application of associated statistical methods statistical quality control and process capacity studies. The vaccination process is described as a random variable with binomial distribution, analyzing its behavior on the application of these vaccines through an np control chart, to quantify the number of wasted doses with respect to the total of applied and a capacity study of binomial processes to determine the quality of this specific activity through sigma levels as established in the 6-sigma methodology. The study included the application period from January 19 to July 11, 2021, following up on the Official Reports and Reports of the Secretary of Health and obtaining the database from their respective portals.

Keywords: COVID-19 vaccination · Quality nursing application · Statistical quality control · Sigma level · Mexican scheme vaccination · Vaccination application waste

1 Introduction

1.1 An Introduction

SARS-CoV-2 and the COVID-19 Pandemic

The Causal Agent

The virus is known as SARS-CoV-2, a new beta-coronavirus of the *Coronaviridae* family, Named for its spherical lipo-protein capsule surrounded by multiple spicules (S-glycoproteins) which give it the appearance of a crown. The genetic material in its interior is a single chain of ribonucleic acid (RNA) positive-sense genetic proximity to

F. Gorski et al. (Eds.): MANUFACTURING 2022, LNME, pp. 136–145, 2022.
https://doi.org/10.1007/978-3-030-99769-4_11

two coronaviruses presents in bats ago highly probable that this is its origin, with the possible participation of one or more intermediate hosts [1].

The virus usually enters through the respiratory route (even though it can do so through the mucous membranes such as the conjunctiva), and is fixed by the spicules to its receptor: the membrane protein angiotensin-converting enzyme type 2 (ACE-2), of type II epithelial and alveolar cells. Once internalized, the RNA is released for transcription and replication.

From the first cases reported in Wuhan, China, the local and finally global expansion quickly reached alarming levels. By January 11, 2020 had already claimed the first fatality and a month later on February 11, the Control Center and China Disease Prevention had 72,314 reported cases14. On January 30, 2020, the WHO declared COVID-19 a "public health emergency of international scope", and on March 11 with 37,364 cases reported outside of China, cataloged officially as a "pandemic". For the first time in history, this disease has been followed in "real time" through the different digital platforms, with statistics and detailed data day by day and minute by minute.

Nursing and COVID-19

Pandemic Environment

For Mexicans, the strategy called "Operation Correcaminos", whose purpose is to achieve coverage effectively and efficiently, with the general coordination being the president. Andrés Manuel López Obrador and the 32 sub-coordinators state. 10 thousand vaccination points are considered in the 32 states of the Mexican Republic with brigades of 12 members (health personnel, SEDENA, Navy, servants of the nation attached to the Ministry of Welfare and volunteer staff if necessary) [2].

During the pandemic there have been many contributions that the nursing professional has offered to the health system, one of the most important is to make visible the need to increase nursing human resources, which has historically had a significant deficit and which the pandemic has translated into an imperative need to recruit health personnel with the intention of forming the "covid teams" indispensable to care for the population. This pandemic has redirected the gaze of the population and health authorities towards the nursing profession, so that it is now an essential figure that has ceased to be only part of the multidisciplinary team, to become a protagonist in the health scene, since it has responded immediately and forcefully to the needs to interact for the well-being of society and throughout the world it has been a fundamental pillar to preserve the lives of those who have required it.

The nursing professional's response is the result of their dedication, dedication and vocation. It is worth mentioning that along the way there have been major obstacles which have undoubtedly hindered action in the various settings in which one travels as a worker in a health institution or as a normal citizen. These obstacles have not limited their practice, on the contrary, they have strengthened and empowered the profession, always courageously despite long working hours has spoken out to face up to the challenge care of people with the objective of ensuring security and providing personalized accompaniment, attending to and prioritizing their biological, psychological, social and spiritual needs, using all available resources. It is also extremely important to reflect on the risks nurses face on a daily basis in the workplace, where fear and lack of therapeutic and non-therapeutic inputs have predominated, such as personal protective equipment, stress, emotional burdens, fatigue, isolation without family living together, and attacks on health personnel. Demonstrations to protest for their rights, etc., as well as solidarity, teamwork, the development of new nursing skills, constancy, discipline, adaptation, resilience and many other positive aspects that have helped save thousands of lives have also been relevant. There is no doubt that this pandemic will continue to generate fear of contagion and uncertainty due to occupational risk, although today lifestyles and routines have changed, the mysticism of the nursing profession has placed it in the perspective of an independent and necessary profession in the world, so it is worthwhile to continue working to have a healthy population and a profession with presence and recognition before society. The intervention of nursing professionals in the covid-19 pandemic has been an unprecedented test that will contribute to knowledge, which has highlighted the importance of providing humanized care to patients who have required it, since their intervention favors integral recovery for a prompt incorporation into the family, work and social nucleus or, failing this, to seek a dignified death [3].

In Mexico, the health system aims to establish instruments that promote the quality and efficiency of health services, expand social security coverage, accelerate and deepen decentralization, and extend service coverage.

The Evaluation of the Quality of Health Care is a first step for the Reform of the System in poor urban and rural areas, through a basic package of services [4]. In 1999, the National Health System of the United States (NHS) published the so-called "Framework for Performance Evaluation" in which the need for actions for its quantification in six areas is indicated, presenting a series of indicators for each one.

On the other hand, high demand leads countries to work with high quality indices to reduce waste as much as possible and increase the capacity of their operations to meet the global demand for vaccines.

1.2 Statistical Quality Control

About Statistical Quality Control

Statistical quality control is a technology widely used in manufacturing to improve product quality and worker productivity (Wetherill 1977). The technology involves first describing the average and the normal variability of work performance outcomes. Repeated samples of work outcomes are then taken and compared to these established standards. Normal and abnormal variation among samples is visually determined with a control chart. The expected performance standard and ranges of expected variation in samples are illustrated on this control chart. Sample data are graphed on the chart and evaluations made on whether the process is in control (normal variation) or out of control (abnormal variation). If the work sample is out of control (i.e., three standard deviations from the average), the variation is abnormal and steps should be taken to identify and solve the problem at the sources of the abnormal variation.

Statistical quality control is most applicable in situations in which a high-rate work activity produces a measurable outcome, for instance, in assembly line work. This situation enables managers to frequently sample work quality. One advantage of frequent sampling is that the average and the variability of the work performance can be quickly established and, another, that worker productivity can be easily sampled. Thus, problems with the quality of the work system can be identified on a timely basis and before the system has been out of control for too long.

These latter advantages have led statistical quality control to be more widely used in manufacturing than in the service or health care industries has discussed the application of statistical quality control technology in developing and interpreting key indicators of patient care, no other examples of statistical quality control applications in the long-term health care field have been published. Furthermore, there appear to be several areas of patient care in nursing homes that could be monitored with a powerful statistical quality control technology. If a patient care area that requires nursing staff to engage in a repetitive work activity has established clear outcome measures, then statistical quality control can be used to determine the quality of this work activity [5].

The control chart, is the main instrument of statistical quality control to analyze the variation in the long term [6], they allow to continuously monitor the accumulated differences of an attribute which is defined as any aspect of the product that by itself can be used to compare others of a given process [7], which can be translated into data collected sequentially, allowing early detection of deviations from an established standard.

One of the advantages of these sequential tests is the independence of the sample size, a greater power to detect transitory changes in trends, the continuity of the analysis over time and the possibility of making a rapid evaluation of the data and a timely identification of the data. behaviors.

1.3 Six Sigma Methodology

About Six Sigma Methodology

Six sigma methodology is a group of techniques and tools which is used for improvement in the process. Six sigma was firstly introduced in 1986 by Mr. Bill Smith and Mikel J Harry when they were working with Motorola Company. In the year of 1995 Jack Welch use six sigma for his business program. It is used for quality improvement of the process and process output is identify and remove the causes of defects and minimize variability in manufacturing and business program. Six sigma aiming at the reduction of defect rate to 3.4 defects for every million opportunities. Six sigma as a project based methodology for solving specific performance problems recognized by an organization. Doing things in best possible way and keeping it in right direction by six sigma. Kaushik gives a definition for six sigma "methodology that offers reliability and giving approach to solve the problem by team and a management system that helps in making leadership and give authority for problem solving in industry" [8].

Six sigma projects methodology: Six sigma projects follow two methodologies. These methodologies are DMAIC and DMADV. DMAIC: - Aim of this process is improving an existing business process. DMADV: - Creating new product or process designs by this process. In this paper we will discuss about DMAIC methodology:

#Define: In this phase consider, voice of the customer and about their requirements, and define goals of a projects. #Measure: In this phase measure gauge repeatability & reproducibility of the running process and check the process capability of a project.

#Analyze: Data is collected and develop a flow of process to analyze and verify cause-and effect of a process and what is the root cause of this defect.

#Improve: Improve the running process based upon data analysis using techniques such as DOE, FMEA, Pareto chart is used for improvement.

#Control: Standardize and documented the improvement of the process control chart is a tool which is used in this phase to check the process problem is shift or not [9].

1.4 Vaccination Process

Official Procedures

The Mexican government issued 7 technical guidelines for the application of the vaccines: BNT162b2 Pfizer/NioNTech, Janssen Recombinant Linear Vector, GAM-COVID-VAC (Sputnik V), Spikevax (Moderna), recombinant vaccine against the new coronavirus "Vector of Adenovirus type 5" (CanSino), Astra Zeneca, Sinovac vaccine "SARS-CoV-2 (Vero Cells) inactivated" all against the SARS-CoV-2 virus and the Operational Vaccination Strategy against COVID-19 "Operativo Correcaminos" [10].

SARS-CoV-2 virus vaccine application technique

- Prepare vaccine.
- Discover the application site.
- Remove the protective sheath or cap from the needle to give the vaccine.
- With one hand stretch the skin.
- With the other hand, hold the syringe, with the bevel of the needle facing upwards at a 90° angle above the plane of the skin.
- Insert the needle intramus cularly.
- Press the plunger so that the vaccine penetrates.
- Lightly fix the skin with a dry swab, near the site where the needle is inserted, and remove the syringe immediately after inserting the liquid.
- Stretch the skin to lose the light from the hole left by the needle.
- Press for 30 to 60 s with a swab, without massaging.
- At the end of the procedure, perform hand hygiene.

Defects on Vaccination Technique

During the vaccination event there are two possibilities, one of success and the other of failure or Defect, in this paper we will talk about "defect" as those doses that were wasted during the preparation or application process.

The vaccination process is described as a random variable with binomial distribution, analyzing its behavior on the application of these vaccines through an np control graph, to quantify the number of wasted doses with respect to the total of applied ones.

A binomial process capacity study is carried out to determine the quality of this specific activity through sigma levels as established in the Six Sigma Methodology.

1.5 Motivation

Currently, the management of the application of the COVID-19 vaccination does not integrate a quality dimension as part of the success of Vaccination Programs in the world. It is important for the World Health Systems to communicate the efforts of its Operational Strategy, not only in terms of the number of people vaccinated, but also the quality with which the process is being executed in order to deploy concepts that They are not always within the reach of the entire population and unfortunately these types of events do occur.

1.6 Objective

Propose statistical quality control as a decision-making tool within the vaccination system to achieve high levels of quality. The hypothesis of the study is centered is to determine if the vaccination scheme presents a rate of 0.10% of defects and is in control.

2 Methodology

This work is based on the study of the quality of the application process for COVID-19 vaccine in Mexico, during the period from January 19 to July 11, 2021, through the application of statistical methods associated with statistical quality control and process capacity studies.

$$P(y) = P\binom{n}{y}p^{y}q^{n-y} \tag{1}$$

The *np* graph is also based on the Binomial distribution. So it is possible to return to some definitions and formulas, it is known that:

$$\mu = E(Y) = np; \sigma 2 = V(Y) = npq; q = 1 - p \tag{2}$$

Let $X_1, X_3, \ldots, X_k$ be a set of samples drawn in the sampling interval i = 1, 2,..., k of a process, denote with *ni* the sample size, and let *n*, $Y_2, \ldots, Y_k$, the number of defective units in the sample. Then, *Yi* is binomially distributed with parameters *n* and *p*.

For each sample, the random variable "proportion of sample defectives" is defined as in the expression:

$$\hat{p}i = \frac{Y_i}{n}$$

The mean and standard deviation of *pi* is:

$$\left(E\hat{p}i\right) = \frac{E(Y_i)}{n} + \frac{np}{n} = p \tag{3}$$

$$V\left(\hat{p}i\right) = \frac{V(Y_i)}{n^2} = \frac{p(1-p)}{n} \tag{4}$$

A random variable Y is said to have a binomial distribution based on n trials with probability p of success if and only if:

With the use of these parameters, the control limits are derived, and they are as follows:

$$\mathrm{UCL} = np + Z\frac{\alpha}{2}\sqrt{np(1-p)} \tag{5}$$

$$\mathrm{LCL} = np - Z\frac{\alpha}{2}\sqrt{np(1-p)} \tag{6}$$

$$CL = np \tag{7}$$

Process capability can be described in units of sigma; For example, a process can be located at a 4.8 sigma level of a Six Sigma ideal, this percentage that is out of specifications can be considered as the defective fraction of the control charts p and np, while the number of sigma approaches Six Sigma, the process will be at zero defects.

In the case of performance metrics, these are based on the products that come out compliant or non-defective, this metric is represented in percentage terms and is also called the compliant fraction of the process.

The equations for assessing metrics, such as Defects in Parts Per Million Opportunities (DPMO), Performance (Y), and Sigma Level (Z), are presented below.

$$\boldsymbol{DPM}0\frac{\boldsymbol{n}}{\boldsymbol{T}}(1,000,000) = \frac{\boldsymbol{n}}{\boldsymbol{UxO}}(1000,000) \tag{8}$$

where n = defects, u = sample, o = error opportunities.

$$\boldsymbol{Y} = \left(1 - \frac{\boldsymbol{n}}{\boldsymbol{UxO}}\right) \tag{9}$$

where n = defects, u = sample, o = error opportunities, y = yield.

$$(\mathrm{NORMSINV}(1 - \%\mathrm{ERROR\ RATE})) + 1.5 \tag{10}$$

All the information on the applied and wasted vaccines were collected from the official site of the Mexican Secretary of Health [11] and where they were processed with the Minitab19 ® software.

3 Results

To monitor the behavior of the discarded vaccines, an *np* graph (Fig. 1) and the binomial capacity study (Fig. 2) were constructed using Minitab 19 ®.

In Fig. 1, the behaviors of 163 days of monitoring are observed, which show at the beginning very low levels of dose wastage from January to the end of March, when they begin to increase by 500% until an average of 404 doses wasted until the first week of July.

On the other hand, in Fig. 3, it can be seen that the performance of the application of the COVID-19 vaccine in Mexico presents a waste of 0.12%, 0.02% higher than the institutionally estimated, statistically not significant (a test of proportions was used t-student).

The measurement of Quality in the application of the vaccine measured in sigma levels was estimated at 4.54 sigma and with the defect rate of 404 wasted vaccines per day out of a total of 50,698,518 applied until July 11.

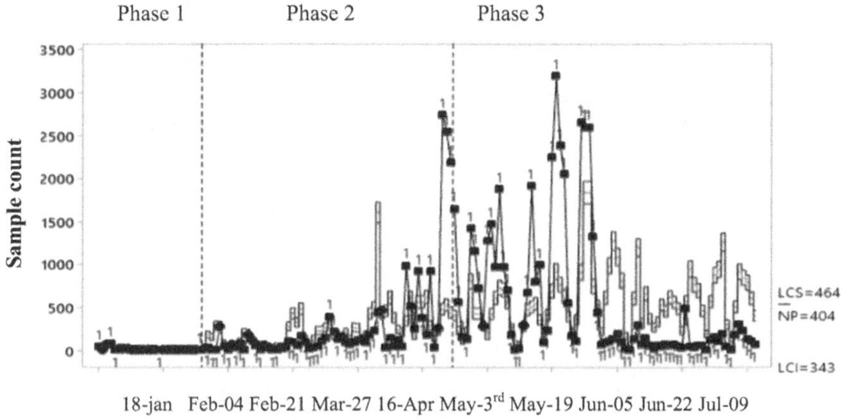

Fig. 1. *np* Chart behavior of the application of the antiCovid vaccine in Mexico (January–July 2021).

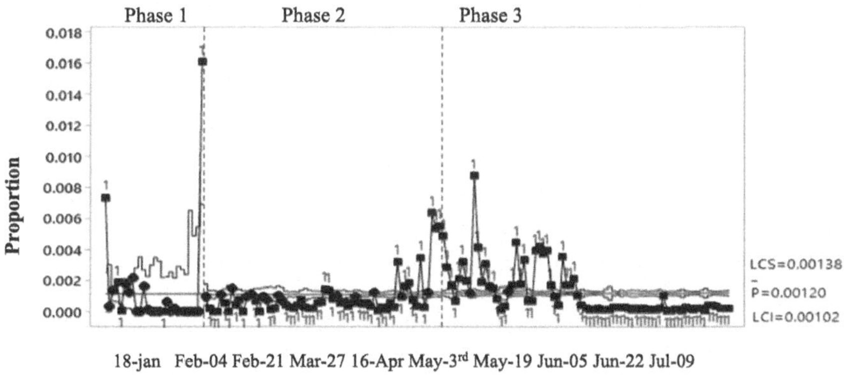

Fig. 2. *P* Chart behavior of the application of the Covid-19 vaccine in Mexico (January–July 2021).

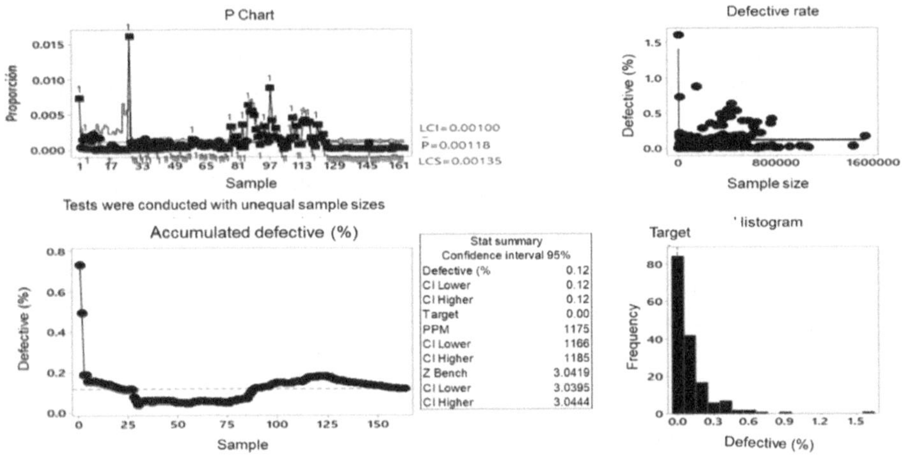

Fig. 3. Study of the capacity of the Covid-19 vaccine in Mexico (January–July 2021).

4 Conclusions

The way in which technology has advanced, has allowed data to be collected with these 4 characteristics: speed, volume, variety and veracity, which are present in an event of high impact on humanity as a pandemic. These advances allow not only to capture events, but are also vitally important in decision-making for every government in the world, which allows you to act preventively in the creation and improvement of their strategies to deal with this virus COVID-19.

One of the disciplines that support the translation of data into information is Statistics, and within it methods, such as Statistical Quality Control, which has been widely used in the improvement processes at the industrial level and in the service area for many years around the world.

On the other hand, the concept of Quality is disseminated in the media and as part of marketing campaigns in the promotion of brands, but in a subjective way providing an idea about the experience that the consumer will have when purchasing a product or a service, which is useful during the purchasing decision process.

With regard to this pandemic event, the Mexican government held a series of conferences to inform the public about the progress of the epidemiological strategy and the vaccination plan. In this forum, information is shared on deaths, contagions, doses applied and wasted doses, where the latter are regularly mentioned as insignificant, but which are not important. at the same time represent the entire effort of thousands of health professionals involved in the direct process of vaccine implementation.

The importance of knowing the behavior of the application in the population defined as applied and wasted doses is vital in the process of awareness-raising and of course the extra cost involved in each wasted dose, which we call Quality in the context of processes. Being able to understand in numbers an entire operational strategy helps the population to appreciate the efforts of the government regardless of their political party, to recognize the work of the entire operational chain and of the personnel directly linked to the fulfillment of the activities, such as: nurses, soldiers, transporters, doctors, volunteers, cleaning staff, planning, among many other people we don't see.

On the other hand, the Mexican State should be more interested in the quality of its processes, the use of tools and methods that support it in carrying out its functions with world-class standards at the lowest possible cost. Although it is not relevant for it to speak of a portion of wasted vaccines, for people on the battlefield it can be a great motivator to recognize their In the same way, people will be able to perceive the concept much better as part of the process of transformation that they themselves have commented on in their political discourse, as a fundamental basis for the economic development of the country.

This small research represents a small tribute to all those people who have worked, even died, to offer their knowledge saving lives during the COVID-19 pandemic in Mexico and around the world.

The study shows how through statistical quality control and the use of the *np* chart and the use of graphical capacity studies, possible behaviors associated with fatigue or ergonomic problems that intervene in the increase in defect rates can be studied. This type of study is of vital importance since it allows us to evaluate the design of the programs, make decisions in a timely manner, identify behaviors that warn of errors, maximize the use of doses and build an autonomous system within the vaccination scheme.

References

1. Escudero, X., Guarner, J., Galindo-Fraga, A., Escudero-Salamanca, M., Marco, A.: Alcocer-Gamba y Carlos Del-Río La pandemia de Coronavirus SARS-CoV-2 (COVID-19): Situación actual e implicaciones para México. Arch Cardiol Mex **90**(Suppl.), 7–13 (2020). (in Spanish)
2. Orellana, C.J.E., Guerrero, S.R.N.: El proceso de vacunación en México. Rev. ADM **78**(5), 270–274 (2021). (in Spanish)
3. Zárate Grajales, R.A., Ostiguín Meléndez, R.M., Castro, A.R., Valencia Castillo (Compilers), F.B.: Enfermería y COVID-19: la voz de sus protagonistas. Universidad Nacional Autónoma de México **2020**, 36–39 (2020). (in Spanish)
4. Jiménez, E.R.: Indicators of quality and efficiency of hospital services. A current look. Cuban Public Health Mag. **30**(1), 17–36 (2004)
5. Llinas, D.A.: Evaluation of the quality of health care, a first step for the reform of the Salud. Uninorte Syst. **26**(1), 143–154 (2010)
6. Schnelle, J.F., Newman, D.R., Fogarty, T.: Statistical quality control in nursing homes: assessment and management of chronic urinary incontinence. Health Serv. Res. **25**, 4 (1990)
7. Noskievičová, D.: Complex control chart interpretation. Int. J. Eng. Bus. Manag. **5** (2013). https://doi.org/10.5772/56441
8. Pohlmeyer, A.E.: Identifying attribute importance in early product development. Exemplified by interactive technologies and age. Ph.D. thesis, Technische Universität Berlin (2012)
9. Stephan, J., et al.: Using six sigma methodology to reduce patient transfer times from floor to critical-care beds. J. Healthc. Qual. **34**(1), 44–54 (2012)
10. http://vacunacovid.gob.mx/wordpress/documentos-de-consulta/
11. https://www.gob.mx/salud/documentos/presentaciones-de-las-conferencias-de-prensa-2021

Quality Assessment of the Cross-Linking Process of Vascular Prostheses

Agnieszka Kujawińska(✉), Michał Rogalewicz, and Joanna Pohl

Poznan University of Technology, Piotrowo 3 Street, 61255 Poznan, Poland
agnieszka.kujawinska@put.poznan.pl

Abstract. The object of research was the process of manufacturing of four types of synthetic vascular prostheses: straight, bifurcated (Y-shaped), straight with an arm and arch-shaped. The main purpose of the paper was to evaluate a quality assessment of this process and as it turned out to improve its quality. The main critical features affecting the quality and durability of the prosthesis are water, gelatin and glycerin. Because of a high cost of examination, control plan designed for this process assumed collecting a very small sample. Taking into account a high risk of accepting a batch that does not meet the requirements, distributions shape and capability for chosen features were assessed on the basis of previously collected data. The content of water turned out to be critical so a thesis was made that the critical operation in the entire prosthesis manufacturing process is the drying operation, both after coating and cross-linking. It takes place in a separate room where wet prostheses are placed for a specified period of time. The company tries to maintain a certain temperature and humidity in the room. Unfortunately, these are not permanent conditions. Therefore, it was decided to analyze the correlation of room temperature and humidity with the content of water, gelatin and glycerin in prostheses. After all, in order to verify the nature of the dependencies obtained in the preliminary tests, a controlled method of carrying out the drying operation after cross-linking and conducting an active experiment was proposed.

Keywords: Vascular prostheses · Process capability · Design of Experiment

1 Introduction

The aim of each manufacturing process is to generate a product that meets customer requirements [1]. It is possible to achieve it, among other things, through manufacturing processes which are statistically stable and qualitatively capable. In order to be able to assess the stability and capability of processes, they must be rationally and effectively monitored and controlled [2, 3].

In a broad sense, control is one of the basic management functions. It is used in all areas of enterprise operation. It consists in comparing the actual state of an object (process, product, document, position) with the assumed (required) state. In order to make this comparison, the state of an object should be [4, 5]:

F. Gorski et al. (Eds.): MANUFACTURING 2022, LNME, pp. 146–160, 2022.
https://doi.org/10.1007/978-3-030-99769-4_12

- measured (through measuring) or described (through observation) to determine the actual state,
- assessed whether the actual state is consistent with the assumed state (or to what extent it is not consistent).

In the production process, control activities occur in product design, in the design of the manufacturing process, in the manufacturing process itself, and after the product has been handed over to the recipient. One of the widely used forms of control is statistical acceptance control [6, 7]. It consists in taking an n-element sample from a batch of finished products and issuing a decision concerning the compliance of the entire batch on the basis of the sample control results. The acceptance control plan takes into account the acceptable quality level agreed between the recipient and supplier, the quantity of the batch presented for control and the level of control. On this basis, a decision is made on the size of the sample submitted for assessment and the size of the batch acceptance criteria [8].

Statistical acceptance control carries the recipient's risk that the batch will be accepted on the basis of a random sample test despite not meeting the requirements and the supplier's risk that the batch delivered will be rejected on the basis of a random sample test despite meeting the requirements. Nevertheless, an important issue in planning acceptance control is the relation with the cost of its implementation. In practice, efforts are often made to minimize control costs by reducing the sample size. The sample size is important as it influences the credibility of assessment - it affects the decision whether a batch of products can be accepted as compliant with the requirements or whether it is entirely recognized as non-compliant. It is common practice to increase the credibility of the acceptance inspection by increasing the sample size and changing the party acceptance criteria. Unfortunately, in a situation where the cost of performing the test is high, which is the case with destructive control, this approach is not applicable. Therefore, practitioners should pay attention to the third possibility of increasing the reliability of acceptance control: process improvement by reducing the variability of the assessed characteristics [9].

2 Synthetic Vascular Prostheses

Vascular prostheses are tubes which replace a fragment of a blood vessel, with a straight or branched course (e.g. Y-shaped, Fig. 1). They are usually made of polyester or polytetrafluoroethylene fibers, which are biologically neutral for the patient's body. Subsequent paragraphs, however, are indented.

Vascular prostheses are mainly used as arterial and venous implants to bypass or replace a narrowed or blocked section of a blood vessel. The walls of the prostheses are notched, which increases their flexibility during medical procedures and the possibility of appropriate length selection during operations.

The fabric for polyester prostheses is mainly produced by weaving - the fibers are intertwined at an angle of 90° (Fig. 2a) and knitting - the fibers form a stitch (Fig. 2b).

Prostheses made of woven material are characterized by negligible surgical permeability, so they can be used without prior sealing. They are mainly used during procedures

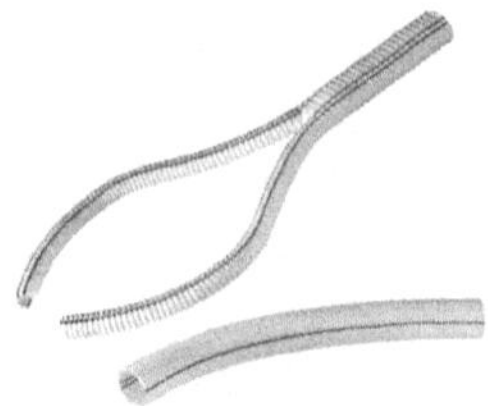

Fig. 1. Example of a shape of prostheses (own source).

to remove thoracic aortic aneurysms and ruptured aneurysms. These prostheses are often used in surgery for patients with blood clotting disorders, with a potential loss of a significant amount of blood (e.g., a ruptured aortic aneurysm) and if they have been given heparin before surgery and before deciding to use a woven implant. On the other hand, knitted prostheses are characterized by porous walls, which increases their permeability. Before using a knitted implant, it should be sealed with the patient's blood, which also ensures better healing. They are used in cases of hip artery surgery and abdominal aorta. Polyester implants can be coated with carbon, silver salts and antibiotics. The use of the previously mentioned chemical compounds for impregnation reduces the possibility of infection and has a positive effect on the prosthesis walls, reducing susceptibility to the formation of mural thrombi.

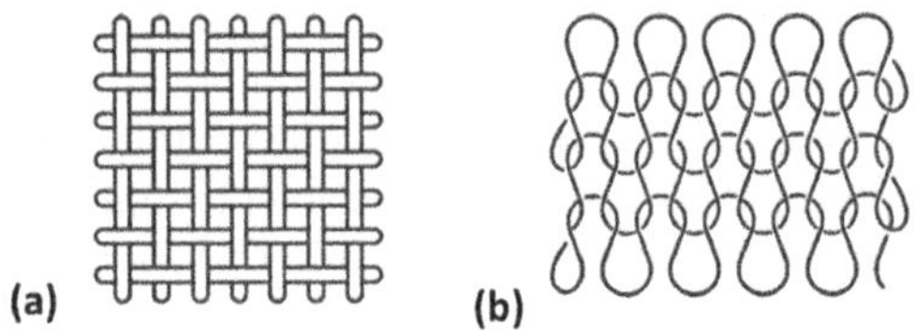

Fig. 2. Diagram of the arrangement of fibers in woven (a) and knitted (b) fabrics (own source).

Polytetrafluoroethylene prostheses are distinguished by an original structure with their walls being uniform, smooth and tight. They are recommended for the operation of small and medium diameter arteries as well as for vein prosthesising. They are mainly implanted in the femoral-popliteal segment. They are also used as shin artery implants and to create an arteriovenous fistula necessary for hemodialysis.

The disadvantage of polytetrafluoroethylene prostheses is their lack of stretchability, which results in the need to precisely assess the required length of the implant. In order to address this inconvenience, a new type of polytetrafluoroethylene prosthesis was developed, referred to as a stretch with a stretchable wall of the prosthesis along its longer axis.

3 Object of Research - Manufacturing Process of Vascular Prostheses

The object of research was the process of manufacturing four types of synthetic vascular prostheses (Fig. 3): straight, bifurcated (Y-shaped), straight with an arm, arch-shaped.

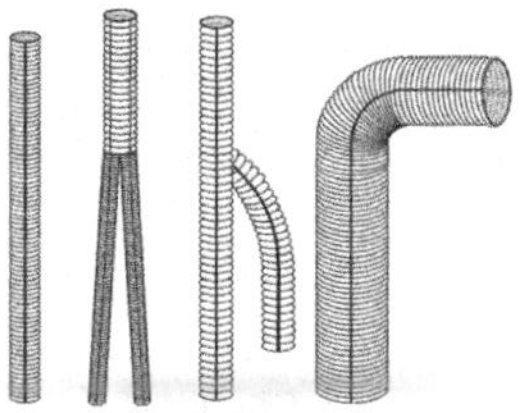

Fig. 3. Types of prostheses: straight, bifurcated, axillo and aortic arch (from left).

For each type of prosthesis, the process is slightly different - the differences lie in the changed sequence of performed activities or different values of the process parameters. In general, the manufacturing process of vascular prostheses always includes the following steps:

Stage 1: Preparation of prostheses: Taking the cut and properly picked semi-finished product from the storage site, which is then subjected to the operation of sewing, washing, shrinking and spinning (the sequence of operations may differ depending on the type of product).
Stage 2: Extraction in isopropanol: In a special column with a capacity of 38 L, a stand with prostheses is placed and treated with isopropanol. The whole process consists of 5 cycles of alternating isopropanol extraction and distillation. When extraction is over, the drying process with compressed air begins automatically.
Stage 3: Notching and pleating: Bearing in mind the group of products, the activities are performed in a different way. The notching process is performed on two devices, while the pleating process is always done manually. The first step is to give the prosthesis a preliminary notch. Each of the prostheses is threaded on a specially adapted form, which is placed in the press, where, under the influence of high temperature and pressure, the fabric is given appropriate size notches. After cooling, the prosthesis is placed on a pleating mold and placed in place by hand. The material is put in an autoclave to give its final shape.
Step 4: Inter-operational control: It is performed before the coating and cross-linking processes and after the coating and cross-linking processes. Interoperative control is a visual control that checks the presence of dirt, stains, contamination and correct pleating.
Step 5: Placing the prostheses for coating: The prostheses are placed on a stand, which will then be placed in a suitable tank. The arrangement of prostheses on a stand (Fig. 4) depends on their type and size.
Step 6: Coating with a mixture of gelatin and glycerol: The process is carried out in two tanks. In the first reactor, a mixture of water, gelatin and glycerin is formed, which is needed to carry out the process. The mixture is then degassed and pumped into the second mixer. Previously prepared stands are placed in the mixture. After 30–45 min, the stand is pulled out and placed in a room where it is dried.
Stage 7: Drying after coating: The prostheses are dried in a special room. The drying time should be at least 15 h in the temperature range: 26–30 °C.
Stage 8: Cross-linking of prostheses: The prostheses are placed for 6 h in a mixture of isopropanol, glycerol, cold WFI and HMDI.

Stage 9: Drying after cross-linking: It takes place in a different room than drying after coating. After drying, the prostheses are inspected, cut to appropriate length, packed and sent for sterilization.

Fig. 4. Adjustable prosthesis stand (own source).

In order to ensure the highest quality of manufactured products and bearing in mind the safety and health of patients who will undergo implantation, a set of key quality features for vascular prostheses was defined. These features are responsible for the permeability, durability and flexibility of prostheses. Many years of R&D tests of products, standards and experimental series made it possible to determine the permissible limit values. As part of quality control carried out in the laboratory, the level of such compounds in the prosthesis as water, gelatin in the form of hydroxyproline, glycerol, ash, HMDI, iron, lead is determined. After cross-linking and sterilization, a certain number of pieces from each batch are sent to an external accredited laboratory.

The main critical features affecting the quality and durability of the prosthesis are water, gelatin and glycerin. The percentage of water in the prosthesis exceeding the upper tolerance line may have an impact on the increased growth of microorganisms on the prosthesis, which is not harmful for the prosthesis itself, but is potentially unfavorable for the patient's health. This can lead to slower wound healing or, at worst, infection. However, if the percentage of water in the prosthesis is below the lower tolerance line, only its physical properties will change. Too little water will affect the flexibility of the prosthesis, but it will not slow down wound healing or increase the risk of infection. When the content of gelatin in the prosthesis is below the tolerance limit, the vascular prosthesis will not be sufficiently flexible. It will bend and leak, which will cause blood to permeate through its surface. Gelatin is soluble in water, and its too low percentage may cause it to be released too quickly from the prosthesis, which must be covered with tissues in order not to lose its tightness. In the other extreme situation, when the percentage of gelatin in the vascular prosthesis is above the upper tolerance limit, the wound may heal too slowly. Gelatin is not a hazardous substance. It is a substance of animal origin, so a significant amount of it will affect health processes. The last critical feature is the amount of glycerin in the prosthesis, which affects its flexibility. If the percentage of glycerin is above the tolerance limit, it does not constitute a dangerous situation for the patient's health. However, the content of glycerin below the lower tolerance line may significantly affect the loss of prosthesis elasticity.

These features significantly determine the quality of the manufactured prostheses. Coating, cross-linking and drying are the main operations shaping the level of water, gelatin and glycerol.

4 Problem

The prosthesis quality assessment process consists of sending several (2 to 4) pieces to an external laboratory. The produced batch is compliant if all the assessed pieces meet the requirements. It turned out that such a control plan was not supported by any risk analysis of the emergence of non-compliant prostheses. Due to the fact that control is carried out on the basis of such a small sample and the results are not related to the range of natural variability of the analyzed statistics but only compared with the tolerance lines, there is a risk of obtaining non-compliant products. Therefore, it is important to estimate associated risk. Hence, the first aim of the research was to analyze historical data from a laboratory for a given type of prosthesis manufactured in 2020. It turned out that the most frequently ordered type of vascular prostheses in 2020 were UG-K prostheses (for which the test results are presented, and the requirements for their selected parameters are presented in Table 1). From each order, 3 pieces go to a laboratory where chemical tests (first piece), tightness (second piece) and the presence of endotoxins (third piece) are carried out. The tests are destructive.

Table 1. Requirements for selected parameters for the UG-K prosthesis (source: company data)

Parameter	Lower tolerance limit	Upper tolerance limit
Water [%]	3	8
Gelatin [%]	13	25
Glycerol [%]	15	30
Ash [%]	0	1
HMDI [ppm]	0	0.1
Iron (ppm)	0	25
Lead [ppm]	0	10

The results of tests from 49 orders for the content of water, gelatin and glycerol were analyzed (Figs. 5, 6, and 7).

The analysis of the test results for the content of water shows that they are close to the upper tolerance line (Fig. 5). These values were obtained from testing one piece from the entire order. This feature is also characterized by high variability, which increases a real risk that the entire batch will contain prostheses with a content above the upper tolerance line, which, as already noted, significantly affects the post-operative healing process and poses a risk of infection.

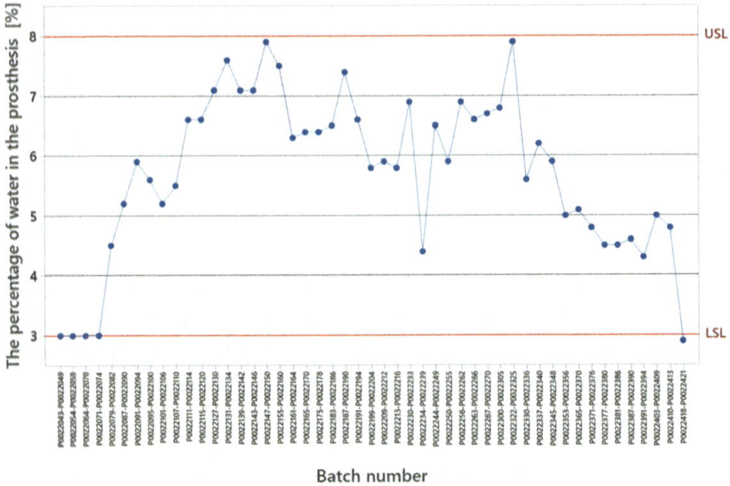

Fig. 5. Graph of the percentage of water in prostheses from a given order (own study).

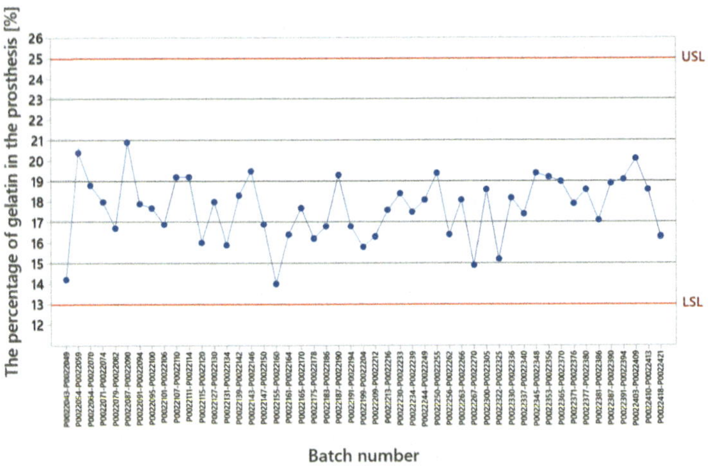

Fig. 6. Graph of the percentage of gelatin in prostheses from a given order (own study).

In the case of gelatin and glycerin content, the distribution of the measurement results from laboratory tests indicates a relatively stable arrangement (Figs. 6, and 7). Nevertheless, the variability of these results also appears to be alarmingly high. This is also indicated by the descriptive statistics from the data presented in Table 2.

The analysis of the test results leads to a thesis that control planned in this way may not reflect the real quality of the order. It is especially visible after performing the qualitative analysis of the manufacturing process (Fig. 8). For the water content, the long-term qualitative capability of the process is very low (Pp = 0.62; Ppk = 0.68). This is information that the potential probability of the appearance of a non-compliant prosthesis is close to 0.05. In the case of glycerol and gelatin, the indices of the quality

Table 2. Requirements for the features of selected parameters for the UG-K prosthesis

Variable	Mean	StDev	Variance	CoefVar	Min	Q1	Median	Q3	Max	IQR
Water	5.720	1.341	1.799	23.45	2.9	4.80	5.9	6.65	7.900	1.850
Gelatin	17.710	1.553	2.411	8.77	14.0	16.55	17.9	18.95	20.900	2.400
Glycerol	21.312	1.908	3.639	8.95	16.9	20.00	21.2	22.65	25.900	2.650

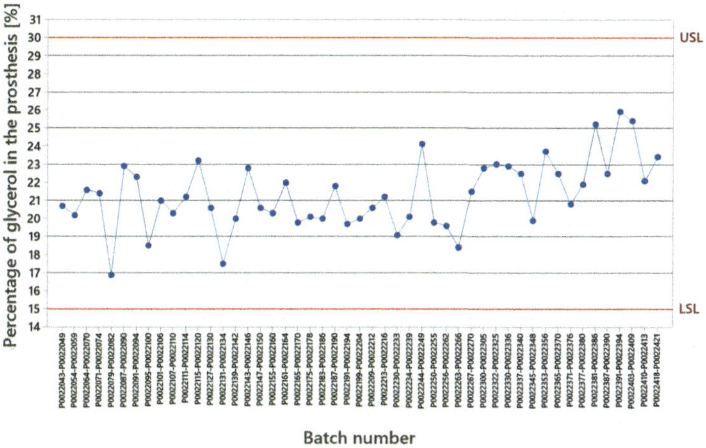

Fig. 7. Graph of the percentage of glycerol in prostheses from a given order (own study).

capability Pp are quite high (close to 1.3), but Ppk is different from Pp which proves a shift of the mean value in relation to the nominal value. It is also not a satisfactory situation.

With such a high risk of non-compliance in the process, the acceptance control plan based on picking one piece carries a high risk of accepting a batch that does not meet the requirements. For example, Fig. 9 shows the Operating Characteristic (OC) curve for a batch of 500 prostheses and the acceptance control plan: sample size equal to 1, $Ac = 0$. From the curve, the probability of receiving the batch at a certain level of non-compliance can be read. It shows that the risk of accepting a batch with a non-compliance level of 4% is as high as 0.96. And with the level of non-compliance: 7% - 0.93.

The conclusion is obvious that the number of measurements should be increased in order to ensure reliable control. In the analyzed case, it turned out to be impossible due to a relatively high cost of producing one piece in relation to the size of the order and the fact that the tests are destructive.

Thus, the only way to improve the effectiveness of the control plan is to minimize process variability. Therefore, it was decided to analyze the process of producing prostheses, in particular the operations in which the content of water, gelatin and glycerin and their tightness are shaped. On the basis of this analysis, a process of their improvement was proposed by performing active experiments which would allow to minimize the variability of critical features, and thus increase the qualitative capability of individual operations.

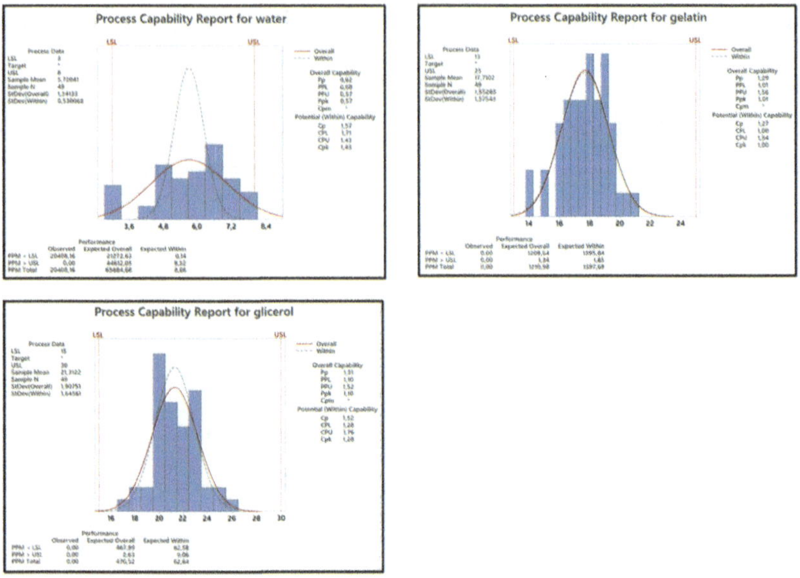

Fig. 8. Indices of long-term and short-term capability for the content of water, gelatin and glycerol in the prosthesis (own study).

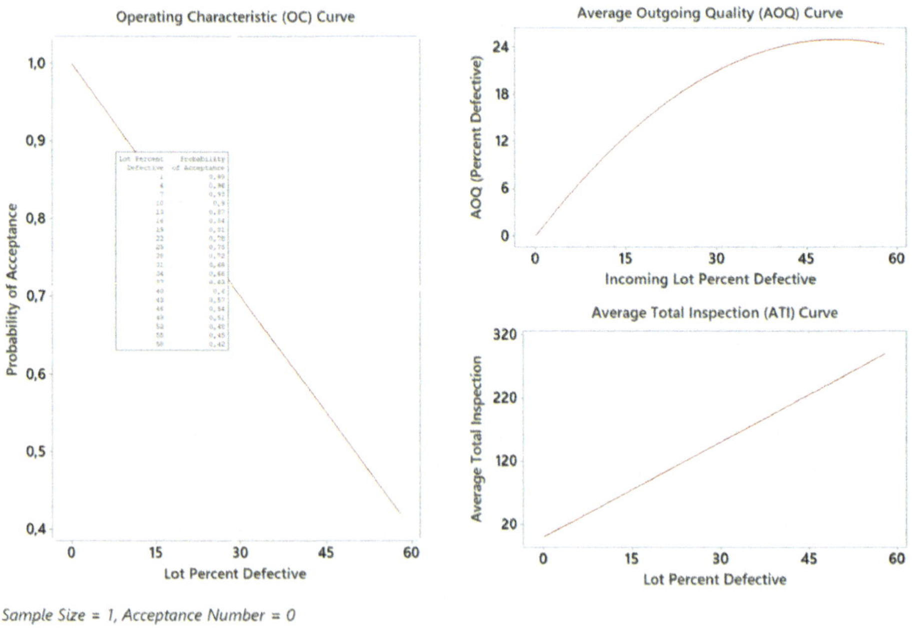

Fig. 9. OC curve for the control plan (N = 500; n = 1; Ac = 0) (own study).

5 Analysis of Factors Affecting the Process

The conducted analysis showed that for all critical features of vascular prostheses, the most variable distribution of measurements was obtained for the percentage of water content. The operations in which water, gelatin and glycerin content are shaped are coating operations, cross-linking operations and drying operations after coating and drying.

In the case of coating and cross-linking operations, their course consists in immersing the prostheses stretched on a stand in a solution prepared according to a proper formula in a special reactor (Fig. 10). A process engineer assessed that the influence of input material factors on these operations is small. The only factors here are: correct preparation of the solution and duration of the operation. These parameters were found to be in line with the applicable standards.

Fig. 10. Tanks used in the coating process (own study).

A thesis was made that the critical operation in the entire prosthesis manufacturing process is the drying operation, both after coating and cross-linking. It takes place in a separate room where wet prostheses are placed for a specified period of time. The company tries to maintain a certain temperature and humidity in the room. Unfortunately, these are not permanent conditions. Therefore, it was decided to analyze the correlation of room temperature and humidity with the content of water, gelatin and glycerin in prostheses.

The analysis of the dependence of ambient humidity and water content in prostheses in the drying process after coating (Fig. 11) allows to conclude that there is a relationship between these features (the value of the linear correlation coefficient $r = 0.531$ is statistically significant). The higher the humidity in the room, the higher the water level in the prosthesis.

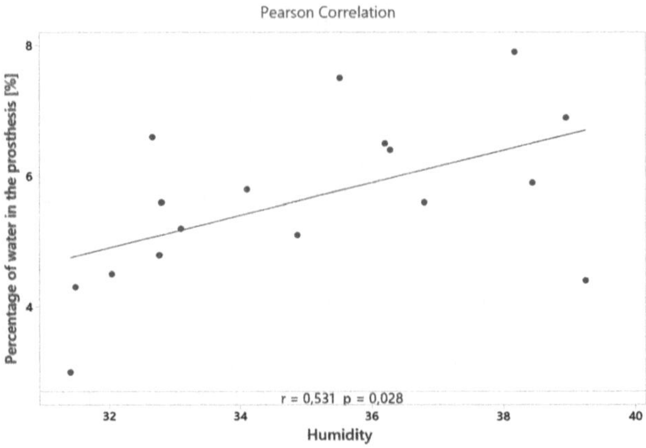

Fig. 11. Diagram of the correlation of the percentage of water and air humidity in the room during the drying of prostheses after coating (own study).

Likewise, the analysis of the relationship between the ambient temperature and water content in prostheses in the drying process after coating (Fig. 12) shows that there is a relationship between them (the value of the linear correlation coefficient r = 0.634 is statistically significant). However, the direction of this correlation seems surprising. It turns out that the higher the temperature, the higher the water level in the prosthesis. This logically contradicting result may indicate the possible existence of another factor contributing to the existence of such a relationship.

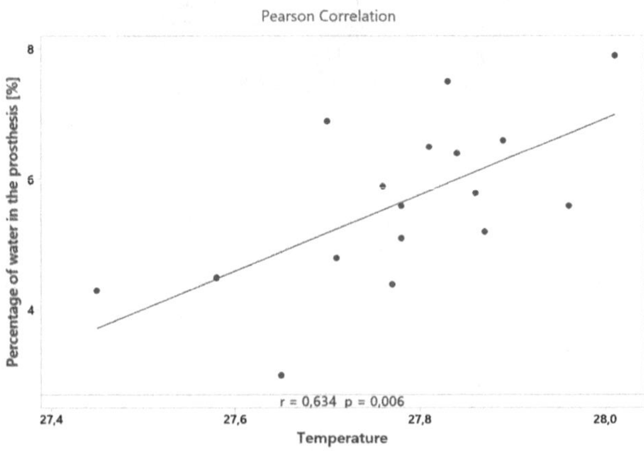

Fig. 12. Diagram of the correlation of the percentage of water and temperature in the room during the drying of prostheses after coating (own study).

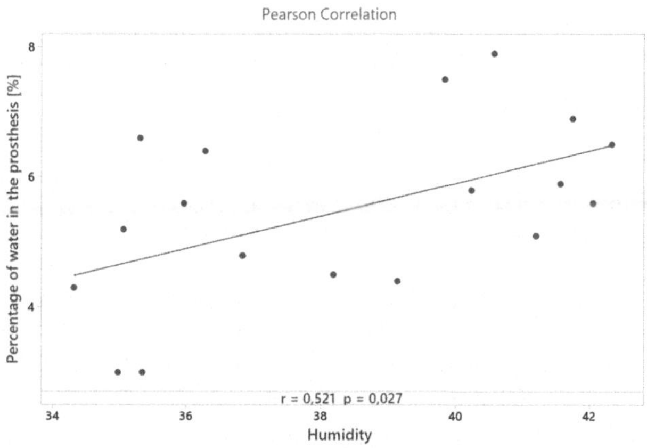

Fig. 13. Diagram of the correlation of the percentage of water and air humidity in the room during the drying of prostheses after cross-linking (own study).

Similar results were obtained in the drying operation after cross-linking (Fig. 13, 14).

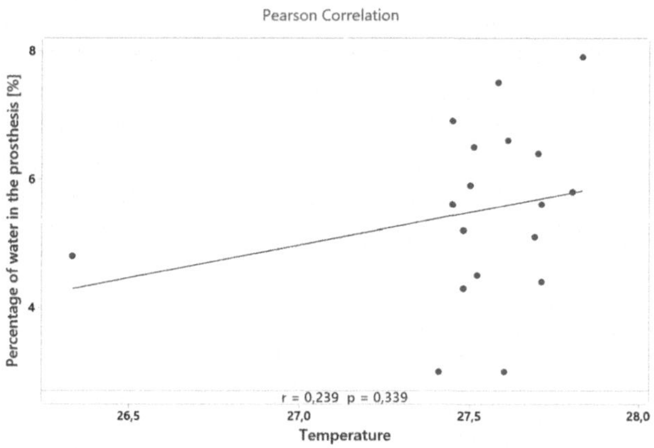

Fig. 14. Diagram of correlation of the percentage of water and temperature in the room during the drying of prostheses after cross-linking (own study).

The results of the analysis make it possible to select room parameters that are not controlled continuously during the process, and which, as one observes, can significantly affect the content of water in the prosthesis, even indirectly. Hence, they were selected as one of the factors for the experiment.

6 Active Experiment Proposal

In order to reduce the variability of the percentage of water content in prostheses, an experiment was planned related to the drying process that most influenced it. As previously mentioned, the drying process occurs twice during the manufacturing process of vascular prostheses. Especially, the drying process after cross-linking determines the chemical values of prostheses because it is the last production stage in which the appropriate properties of the prostheses are given. It is in it that prostheses acquire an impermeable cover and ultimate flexibility.

In order to verify the nature of the dependencies obtained in the preliminary tests, a controlled method of carrying out the drying operation after cross-linking and conducting an active experiment was proposed. For this purpose, it was recommended to use a specialized dryer that would enable an appropriate setting of the analyzed factors. An example of such a dryer is the UN/UF series laboratory dryer with natural/forced convection by Memmert GmbH & Co. KG, which is proposed to be used in the enterprise.

This dryer has the function of regulating temperature, fan speed, program time, humidity, CO_2 and O_2 concentration. It is also characterized by internal data recording, an adjustable clock and the Setpoint WAIT function, which ensures the countdown of time from the moment of reaching the correct temperature. This dryer also guarantees the appropriate size of the working space, which allows drying on specialized stands. In the selected dryer, one should fit 4 stands in one dryer.

In the first stage of the analysis, it was proposed to conduct a full factorial experiment on two levels, which can be further extended, e.g. by an experiment using response surface plans. From a number of potential factors, the experiments selected were those that should have the greatest impact on the final percentage of water content in prostheses. These include the previously analyzed temperature of the air-drying prostheses and the humidity of the air-drying prostheses, as well as a variety of dried prostheses in terms of size. Table 3 shows the low and high values of the factor levels, selected on the basis of the analysis of the preliminary test results and the process requirements. Based on the data from 2020, it is assumed that the proposed intervals should ensure a repeatable production process, thanks to which it would be possible to increase the capability and stability of the conducted processes.

Table 3. Factors analyzed in the factorial experiment

Factor	Description of the factor	Level	
		+1	−1
A	Air temperature	25 °C	27 °C
B	Air humidity	35%	37%
C	Variety of prostheses in terms of size	All are the same	Different

For each conducted experiment, the number of prostheses should be constant. In order to verify the influence of a variety of prostheses on the drying process, a group of prostheses with different dimensions should be selected for the experiment. In the case of an experiment where all prostheses are of the same size, prostheses with a diameter of 8 mm are proposed. In a situation where prostheses are to be different, prostheses with the dimensions of 8 mm, 18 mm, 14 × 7 mm, 16 × 8 mm and 18 × 9 mm are proposed, as these are the most frequently ordered prostheses of the analyzed type. It is recommended that each experiment should be performed in 2 replications in a randomized manner.

7 Summary

The methods of statistical acceptance control are used to check whether the controlled batch of products can be considered as compliant with the requirements. This control is primarily used in the supplier-customer relationship. The parties to the transaction agree on the acceptable defectiveness of the delivered batch and control on the basis of a randomly picked sample is to show whether this defectiveness is not exceeded.

Statistical acceptance procedures are passive quality control methods, i.e. methods used after the completion of technological processes, which do not affect the correct - due to the quality of performance - course of these processes.

It should be emphasized that the assessment of the quality of a batch of products on the basis of the results of the sample quality assessment is always burdened with the possibility of an erroneous conclusion. This error can be of two types:

- type 1 error - consisting in not accepting a batch of products compliant with the requirements (supplier's risk),
- type 2 error - consisting in accepting a non-compliant batch of products (recipient's risk).

The measure of the effectiveness of statistical acceptance is the size of the probability of making these errors. Its value depends not only on the assumed acceptable batch quality level but, as shown in the analyzed example, on the actual batch defectiveness.

The analysis of the qualitative capability of the vascular prosthesis manufacturing process showed that the process is not capable due to the content of water, gelatin and glycerin. Relating these results to the applied acceptance control plan raises doubts as to the reliability of the conclusions of the control results.

With low qualitative capability of the process, there is a high risk of non-compliance in the process. Hence, the acceptance control plan based on picking one piece carries a high risk of accepting a batch that does not meet the requirements.

The analyzed case shows that when it is not possible to increase the sample size, it is absolutely necessary to increase the qualitative capability of the process. In the case of the production of prostheses, the authors proposed to reduce the variability of the content of water, glycerin and gelatin in the prosthesis by changing the drying parameters after coating and cross-linking operations.

In the next stage of their research the authors propose to conduct a full factorial experiment on two levels, which can be further extended, e.g. by an experiment using response surface plans and determine the most favorable conditions of drying process.

Acknowledgments. The paper is prepared and financed by scientific statutory research conducted by Chair of Management and Production Engineering, Faculty of Mechanical Engineering and Management, Poznan University of Technology, Poland, Poland, supported by the Polish Ministry of Science and Higher Education from the financial means in 2021 (0613/SBAD/4710).

References

1. Varela, M.L.R., Putnik, G.D., Manupati, V.K., et al.: Integrated process planning and scheduling in networked manufacturing systems for I4.0: a review and framework proposal. Wireless Netw. **27**, 1587–1599 (2021)
2. Pavlenko, I., et al.: parameter identification of cutting forces in crankshaft grinding using artificial neural networks. Materials **13**(23), 5357 (2020)
3. Ivanov, V., Pavlenko, I., Kuric, I., Kosov, M.: Mathematical modeling and numerical simulation of fixtures for fork-type parts manufacturing. In: Knapčíková, L., Balog, M. (eds.) Industry 4.0: Trends in Management of Intelligent Manufacturing Systems. EICC, pp. 133–142. Springer, Cham (2019). https://doi.org/10.1007/978-3-030-14011-3_12
4. Kujawińska, A., Diering, M., Rogalewicz, M., Żywicki, K., Hetman, Ł: Soft modelling-based methodology of raw material waste estimation. In: Burduk, A., Mazurkiewicz, D. (eds.) ISPEM 2017. AISC, vol. 637, pp. 407–417. Springer, Cham (2018). https://doi.org/10.1007/978-3-319-64465-3_39
5. Bożek, M., Kujawińska, A., Rogalewicz, M., Diering, M., Gościniak, P., Hamrol, A.: Improvement of catheter quality inspection process. In: 8th International Conference on Manufacturing Science and Education – MSE 2017 "Trends in New Industrial Revolution". MATEC Web of Conferences, vol. 121 (2017)
6. Sika, R., Rogalewicz, M.: Demerit control chart as a decision support tool in quality control of ductile cast-iron casting process. In: Proceedings of 8th International Conference on Manufacturing Science and Education (MSE 2017), Trends in New Industrial Revolution, Sibiu, MATEC Web of Conferences, vol. 121, pp. 05007-1–05007-8 (2017)
7. Starzyńska, B., et al.: Requirements elicitation of passengers with reduced mobility for the design of high quality, accessible and inclusive public transport services. Manag. Prod. Eng. Rev. **6**(3), 70–76 (2015)
8. Sika, R., Hajkowski, J.: Synergy of modeling processes in the area of soft and hard modeling. In: Proceedings of 8th International Conference on Manufacturing Science and Education (MSE 2017), Trends in New Industrial Revolution, Sibiu, MATEC Web of Conferences, vol. 121, pp. 04009-1–04009-8 (2017)
9. Starzyńska, B., Klembalska, A.: A digital repository of science assets as a tool for knowledge transfer to manufacturing enterprises. Manag. Prod. Eng. Rev. **12**(2), 115–123 (2021)

A Development Method of a Virtual Reality Environment for Teaching in a Medical Technician School

Leticia Neira-Tovar(✉), Estefania Salisbury Flores, Sergio Ordoñez, Aldo Martinez, and Eduardo Sanchez-Rentería

Universidad Autónoma de Nuevo León, Pedro de Alba S/N, Niños Héroes, San Nicolas de los Garza, Nuevo León, México
{leticia.neira,estefania.salisburyfl,sergio.ordonezgnz, aldo.martinezmr,eduardo.sanchezre}@uanl.edu.mx

Abstract. In the last few years, virtual reality technologies have been applied to training in a variety of field, such as industry or medicine. The use of these technologies aims to promote interaction therefore improving users learning. This approach has been found useful in teaching tasks that involve interaction with risky tools or procedures. Using VR, the user is in a safe environment while developing new abilities required for the field he is learning about. This project is focused on the field of medicine. An immersive virtual reality environment is developed to teach a basic medical procedure to students at a medical technician school, where interactive lessons could be attended in a distance modality. This work aims to be as close as possible to learning in real facilities. Real lessons of the school are translated to virtual exercises where a score is kept to grade performances of students. The lessons involve tasks such as reading a patient's file, practicing hygiene and comfort techniques and muscle strength tests as a health care activity. To increase the sense of presence for the students, the facilities where they would be performing in real life are translated to the rooms of a virtual simulation.

Keywords: Healthcare virtual reality · Medical simulation · Virtual reality training

1 Introduction

Virtual Reality (VR) as an education tool is based on the principle of simulation. This principle is old and common in teaching processes. Simulation is a method or technique that tries to imitate experiences or processes in education, training, where individuals can experience feared situations or contexts, without going through the actual event in real life [1–3]. Mannequin simulation, that is high-reliable in a whole environment, is being used and managed by many medical schools [4]. Nowadays, using the newest technologies, simulation in medical field involves a complete environment allowing the user to interact with them and perform complex procedures such as surgeries [5, 6]

F. Gorski et al. (Eds.): MANUFACTURING 2022, LNME, pp. 161–173, 2022.
https://doi.org/10.1007/978-3-030-99769-4_13

or endotracheal intubations [7]. Simulation is a principle related to teaching medical procedures.

Clinic simulation is a brand-new methodology that recreates a devised setting for experimenting a real event in order to practice, learn, evaluate, prove or acquire a set of cognitive abilities that compromise a better performance in attention to patients, bringing the chance to put their knowledge into the test, to grow confidence in their capabilities by managing different virtual scenarios simulating surgical scenarios, in other words, clinic simulation is an environment created for experimentation [8–10].

Because of globalization and advances in information technology, there is a need to change traditional methods as well as transform the landscape of human-computer interaction with an active new methodology of multisensory and unique perspectives on the core goals of training and education [11, 12]. For that, medical formation has been compared with the traditional model, and teaching based on simulation, where in a clinical essay with real patients, advisor or teachers supervise continuously the alumni to avoid mistakes and they get to correct them instantly because the integrity and security of the patient must be assured. Whereas in simulations, errors are permitted for a better and deeper comprehension in its consequences, rectifying and do the procedure correctly this time, reinforcing their learning skills [13–15].

It is shown that traditional learning does not always guarantee a complete medical profile for graduates. To improve teaching in training and developing professional competences in controlled clinical scenes, medical education programs have been implemented using technologies and software that add up new knowledge and improve quality in attention and efficiency [16]. We can say that simulation in medical education brings a safe and controlled environment, where we can interact with knowledge, challenges, and human factors, it provides an effective way of learning for students. Furthermore, alumni gain knowledge based on reflection to provide an adequate solution to real-life situations in medicine. Because of this, simulation has become a useful tool for improving and learning from mistakes without harming the professional integrity of students [9, 17, 18].

One factor driving its use has been to prevent risks that can be carried out to real patients during hospital practices [19]. Another factor is to avoid direct invasion in the human body, utilizing virtual models for acquirement of new techniques and alternatives on advanced treatments, where because of its complexity and localization, are difficult to access with the current medical equipment [20]. Having an individually programmed learning, acquisition of experience, always having time for analysis and correction of errors and professional help, have promoted the use of clinical simulators in medical teaching [13].

As well as for other fields, the aim of a medical simulator should consider the complexity of the task a user is going to learn, the place where tasks are carried out and the level of experience of the users. According to David Gaba, simulation of medical attention has 11 categories: objectives and purposes of simulation activity; participation unity; participants level of experience, sanitary domain; discipline of participants; type of knowledge; challenges; attitudes or behaviors; age of simulated patient; required or applicable technology; simulation site; degree of direct participation; and used feedback methods [1, 20]. The design of a virtual simulation is based on many factors.

Simulators can be applied to simple or complex procedures. There are specific simulators such as surgical simulators, where doctors train themselves with the latest technology that goes from suture practices to intervention of surgical robots. High-reliable simulators in the specialty of laparoscopic surgery, neurosurgery, endoscopic surgery, trauma surgery, cardiac surgery, and surgical intervention. They contain integrated learning curricula, statistical reports based on performance, such as numbers or movements, and screen recording for evaluation and teacher reporting purposes [8]. However, simulation can be applied to simple procedures such as getting to know the facilities or tools medical personnel will be using. This last approach is applied when there is no previous experience in the field [21]. Either for complex or simple procedures, VR simulations can be applied in medicine. Some elements for user interface are taken from other studies as [22].

The aim of this work is to propose and implement a VR environment for learning medical procedures of a medicine high school to the students. As a technical school, students who finish one of the many programs of the school can later work in the field. In consequence, getting practical knowledge is as important for students as theoretical knowledge. In recent years, the COVID-19 Epidemic has limited access to the real facilities of schools, therefore making it hard for the students to get practical skills. Virtual reality approach can help the students practice those skills from home in a close way to real practice. The simulator proposal includes introduction to basic procedures such as reading a patient's file, protocols for taking care of a patient and validation of muscular strength. VR simulators can help learners to get practical skills from home.

2 Method

The development of the project is distributed in two main stages: conceptualization and design implementation. This method is shown in Fig. 1. This is related by some concepts to Design Thinking (DT), a process to create products, which involves steps as understood, observation, defining the point of view, creating ideas, prototyping, and testing [23]. To create scenes and interactions of virtual environments, development teams need to clarify what should be included as content or view to be a useful tool.

2.1 Conceptualization

The conceptualization stage involves developers getting a general view of the type of projects, the get knowledge from experts related to the specific content of the current project. From background research, a team can identify desirable and undesirable elements in user interaction from another VR simulator applied to the same field. By including the expert experience of someone who knows the real exercises of the target audience, the content will be optimized for the learning of the target audience.

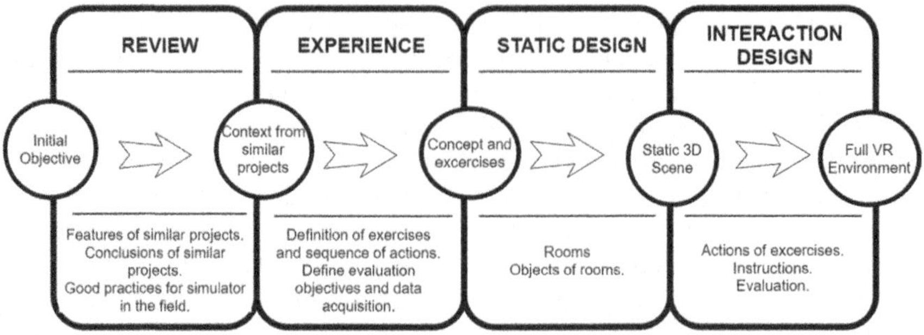

Fig. 1. Methodology used in the development of medical lessons.

Background Research: Development teams must have knowledge of similar projects to the ones they will be carrying out. The use of a research process can be useful to find those related projects. For this case, research to find scientific articles about different medical simulators was made. The research includes database engines as IEEE Xplore, Springer, and Google Scholar. From the found literature, The research focuses in three sections of each paper: design proposal, discussion, and conclusions. In the first section, the team identifies good practices of simulators from previous projects. However, from the other two sections, the teams identify what other teams considered useful for learning or things that should have been implemented in a different way. At the end of the research, the team knows common elements included in this type of VR training tools.

Expert Experience: The experience of expert helps to the simulation content definition. For the training tool to be useful, it needs to be related to the content of the exercises the students would be performing with real training. It also requires giving instructions in a proper way for the audiences expected to use the tool. In this case, the experts that will help with insights into the content are teachers at a medical high school. In this step, sessions between teachers and developers will be carried out. The aim of this session is to establish goals and objectives, and a sequence of actions for each exercise of the training tool. Based on their experience, there is a definition of what the user is supposed to do for getting or practicing a skill.

2.2 Design Implementation

After the conceptualization stage and the meetings with experts, the design process begins. The development team starts creating different sketches of the views, the user interface, the sequence of the simulation and the available user inter- actions [24]. The team divided design into two steps: scene design and interaction design. The first step is related to the view of the VR environment. Once the view was complete, interactions were included following the defined logic of each exercise of the medical high school. Once there is knowledge of the content the VR environment should have, the development team focuses on design tasks.

Scenario Design: For the different exercises of the VR simulator, some rooms and objects were required. The environment should include common rooms of a hospital like the waiting room, a room for patients and an office where patients can be examined. Additional models, including a 3D model of a patient and hospital objects are included. In the case of the design of 3D models are required for each scenario used 3D Max and Blender software. For the design of the icons and backgrounds, Photoshop CC 2018 was used. In this stage of design, views of the environment are created.

Exercises Design: This exercise consists of three practices where different medical procedures and methods for patient care (Fig. 2). The tools used consist of Unity software version 2019. 4. 17f1, in which the C# programming language and Firebase database were to record data from results of students.

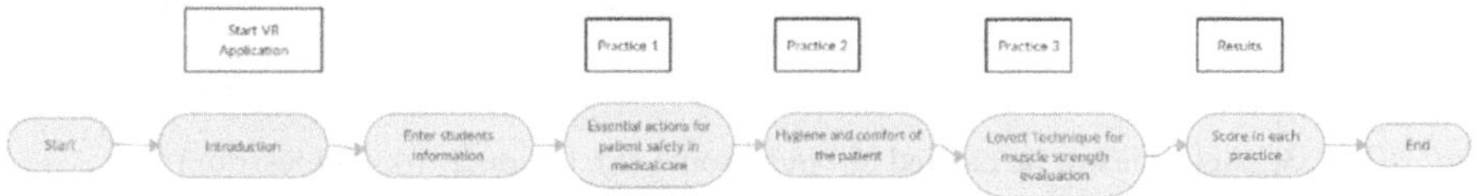

Fig. 2. Flow that occurs in the application from the beginning of the application to the qualification that occurs in all practices.

The aim of the simulator is to record data from students. At the beginning of the practice, you have a main menu with the main options starting the practice. When a user starts, the student must enter his name and license plate. In a database, scores and attempts for each practice are stored. Getting information from students can give information to the teacher of improvement areas.

Each exercise focuses on one aspect of interaction with patients. The first exercise is intended to test essential actions for patient safety in medical care (Fig. 3a). This action involves reading patients file, and identify risk levels, hygiene and security standard protocols, and monitoring actions. Here the primary aim for the student is to ensure the safety of the patients. On the other hand, the second exercise focuses on meeting the needs of hygiene and comfort of the patient (Fig. 3b). This hygiene care also involves reducing bacterial colonization and prevent skin and mucosal lesions. The last exercise of the proposal is related to a specific process: the Lovett Technique for muscle strength evaluation (Fig. 3c). The Lovett Technique involves an initial evaluation by extending the knee, and a student must learn what to evaluate that movement and if there are necessary additional evaluations. The three exercises include different actions that the students learned by real practice.

We used the Acer Mixed Reality Headset equipment to build our simulator. The learner can move and interact with the environment; they can take a history, examine, investigate, diagnose, and treat the patient.

There are additional requirements implemented in design. A web version will be maintained for a wider reach to the student body. This simulator is expected to meet the requirements for recording information and generating reports, in addition to keeping track of and facilitating the search for information. Several addition requirements were also implemented.

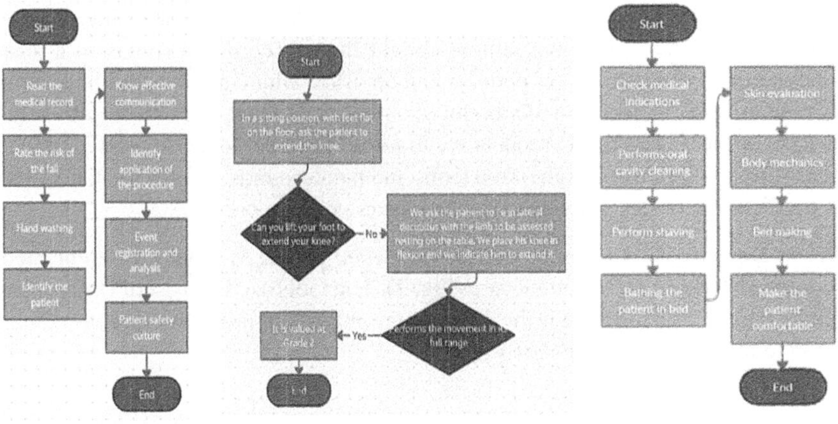

(a) (b) (c)

Fig. 3. Activities to be carried out in lesson 1 (a), lesson 2 (b), and lesson 3 (c).

3 Results: VR Training Tool

3.1 Environment

Three rooms were designed (Fig. 4), one per exercise. During the first exercise, an office with a waiting room was chosen for the students to apply the knowledge acquired, as shown in Fig. 4a. In the second practice, a patient room was used to carry out the actions of hygiene and comfort of the patient (Fig. 4b). Finally for the setting of practice 3, the office of Fig. 4c, where the patient can be examined, was created.

3.2 Exercises

In the first practice, a questioning of the essential actions for the care to be given to patients was conducted. This scene consists of a waiting room where you can see the file that will be seen, as well as a clinic for the patient's check-up to be seen. Figure 5 shows at the bottom the tasks you need to perform, as well as an example of the questionnaire that you perform.

The second practice is performed in the patient's room where the student should be able to verify the medical, oral practice, oral cavity, shaving, bed bathing of the patient, skin assessment, proper use of body mechanics leaves, bed laying and be able to the patient in a comfortable state (Fig. 6). And the last practice is carried out in the office where the muscle assessment is carried out considering the evaluation scale (Fig. 7).

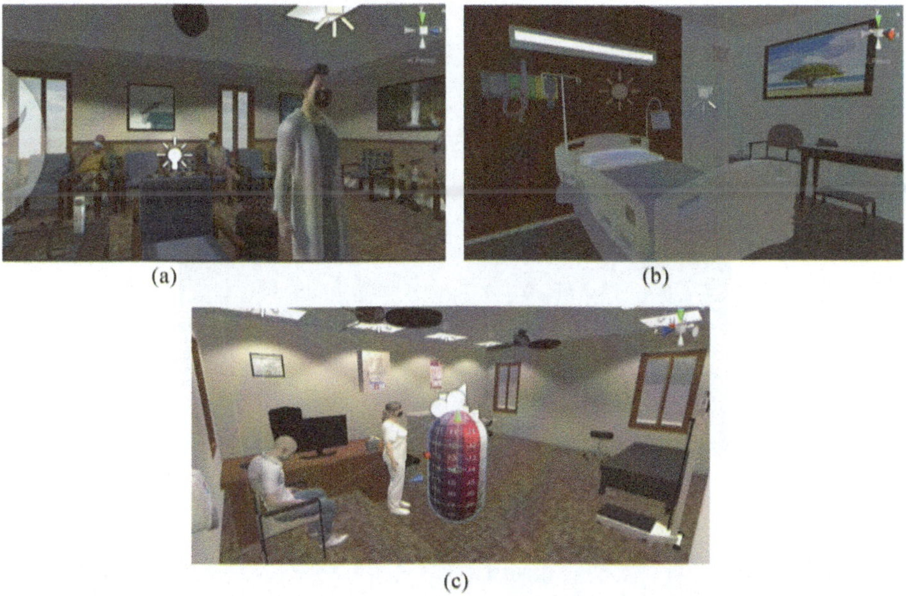

Fig. 4. The scenery was based on a hospital where each lesson takes place in a different department. 3D Representation of different sceneries in the VR Simulator: a) Waiting Room, b) Patient Room, c) Doctor's Office.

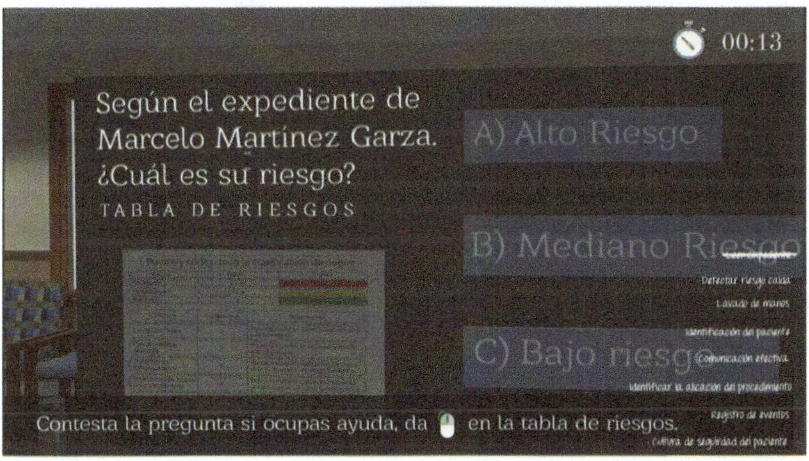

Fig. 5. In this question you must read the patients file and identify risk level and analysis of sentinel events, adverse events, and near-misses.

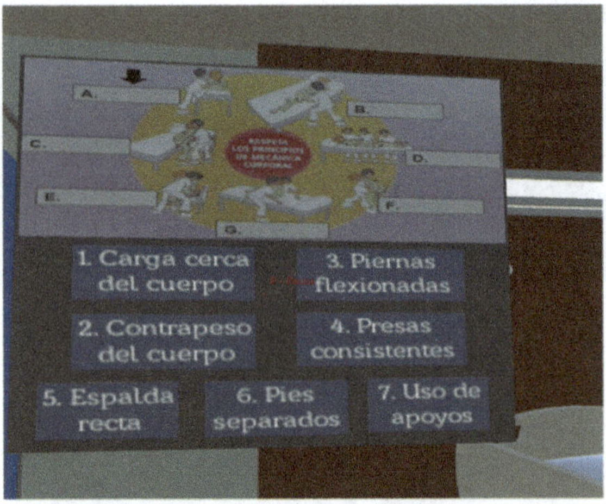

Fig. 6. The student must relate the columns using the principle of body mechanics.

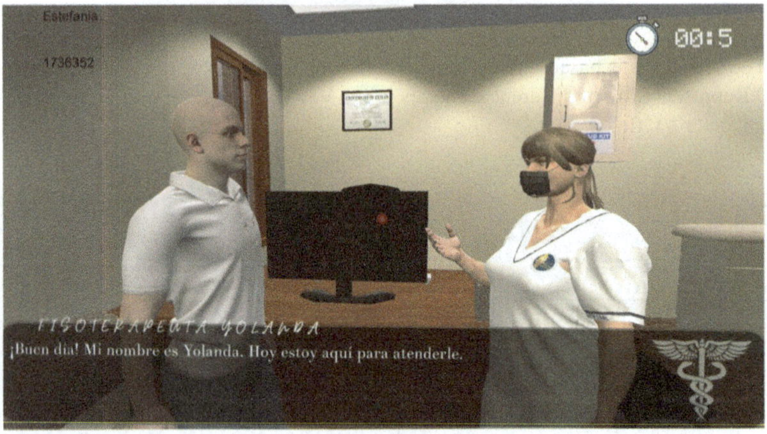

Fig. 7. This lesson reinforces the knowledge of the muscle assessment procedure, considering the muscle assessment scale.

There is data recorded for evaluation. Once the practice has started, a timer and the student's data are displayed at the top of the screen. In the practices, questionnaires were introduced, consisting of selecting the correct order of the steps of the procedure, multiple choice questions, interacting with the characters, memory games and selection of materials to be used. When the practices are completed, a phrase of completion of the practice and the score obtained is displayed along with the number of attempts made (Fig. 8). Students are evaluated in a virtual environment.

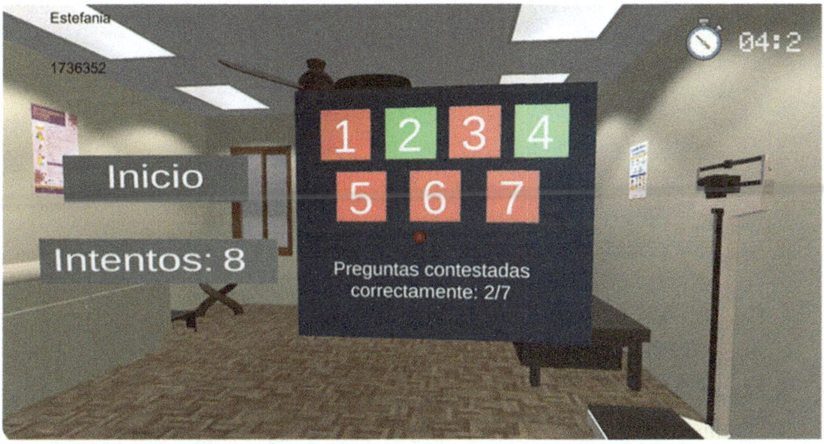

Fig. 8. Score screen: each question is displayed inside a square marked in green, when correct, and red when incorrect. Bottom text shows the score of the practice.

3.3 Experimental Setup

A small group of students participated in a pilot test. A survey was conducted to assess the video game's design and explanation of actions, as well as the effects of learning competency using VR learning methodologies.

The questionnaire covers two categories: feedback and heuristic testing. First category has 13 questions and respondent used Likert scale from one to ten for feedback, where one to ten scale with one implying 'very difficult' and ten being 'very easy'. And in the second category has 15 questions and it is used Likert scale but in this one we use a scale of one to three, where one implying 'strongly disagree' and ten being 'strongly agree'. On average each participant spent fifty minutes to finish all the practices VR experience (Fig. 9).

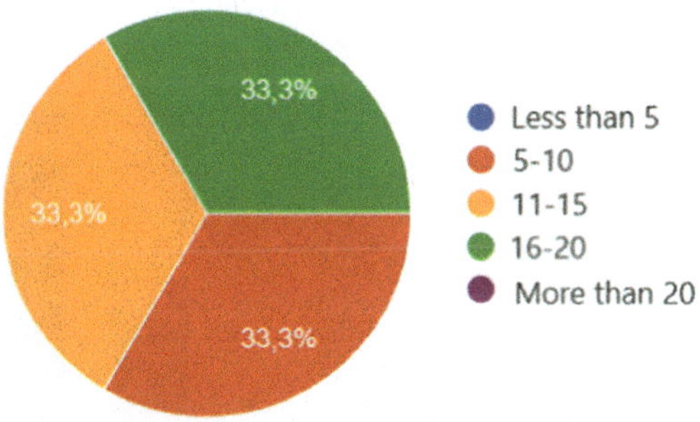

Fig. 9. The amount of time it took students to complete one practice.

Figure 10 shows the bar chart showing the level of agreement regarding achieving practice objectives. In this section, 83.3% of the students strongly agreed the lessons achieve the intended goals. Followed by 16.7% who believe they are barely satisfied.

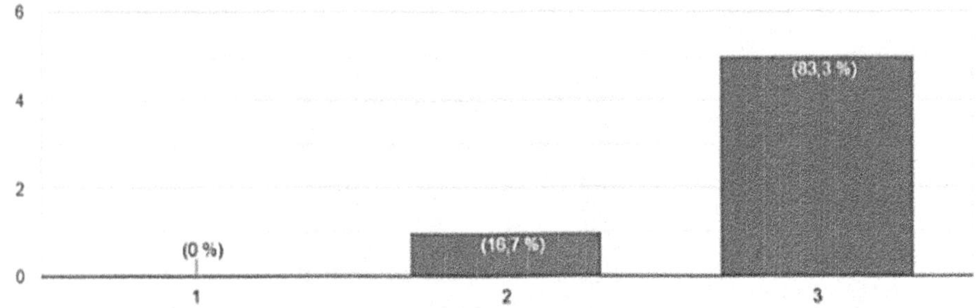

Fig. 10. Bar chart, levels of agreement on achieving practices objectives.

Figure 11 the bar chart depicts the motivation to complete the dynamics of the practice. We can see that 66.7% of the pupils were very driven, while 33.3% simply finished the practice out of obligation.

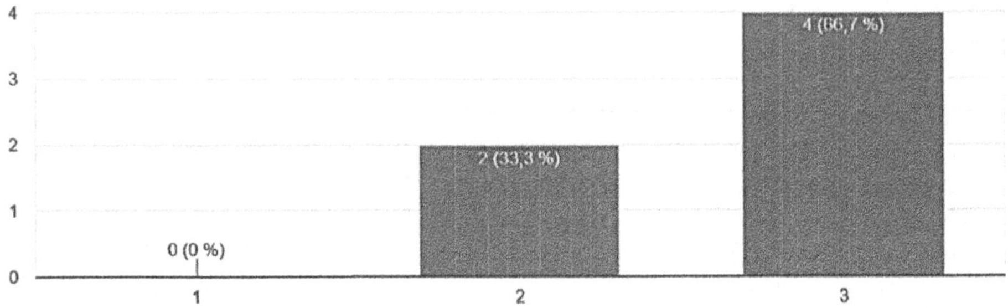

Fig. 11. Bar chart, levels of agreement on motivation for completing the dynamics of practice.

And in the Fig. 12 illustrates the students' enthusiasm for this application 33.3% of students strongly favor this portion. Following that are 16.7% who like it a lot, 33.3% who like it, and 16.7% who simply like the experience.

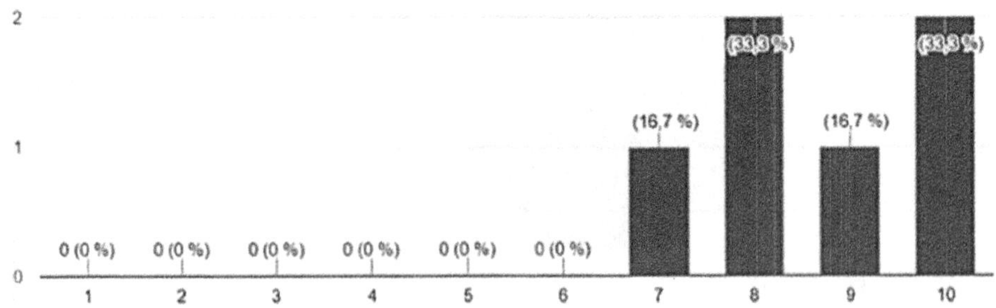

Fig. 12. Bar chart, levels of enthusiasm the pupils were about studying using this innovative way.

4 Discussion

As an environment based on the experience of teachers, this method can be useful to develop tools focus on learning skills of the target audience: the students and the health assistance personal. The exercises implemented in VR are based on real training the students receive when training in medical school facilities. Some suggestions from professors were considered based on their experience training previous groups of students. However, this environment is jet a proposal. Usability testing, as well as learning tests, is required to evaluate how effective the tool is for learning and what improvements are needed. Testing should be considered in the future work of this project. Experience of teachers can help improve a virtual training environment but testing and improvements are needed for tools to be useful.

The proposed environment may be too specific for the target audience. This project was developed in collaboration with medical high school professors to help students learn required field skills from home. The different exercises and the rooms implemented in this proposal are based in detail on the real facilities. If the proposal is proven useful in later work and then applied to other institutions, results may not be as good as in the original school. The concept of each exercise is general and can be used by other institutions. However, static design may need some changes to be more general in the future. If static design is specific, students relate virtual and real world, so it is easier to translate what they learn in a virtual environment to real life. When generic scenarios are used, the learning process should be improved by adding other tools. There are improvements to the proposal design when new objectives arrive.

5 Conclusion

Starting from academic purposes such as surgical simulators, even being able to improve hospital service and medical studies. Although, even knowing all the benefits and advantages that the implementation of these technologies has over other methods, there are still technological barriers because hardware and software are expensive, and expertise is limited. In the future, this will be broken down further [25].

Today, there are many institutions that use VR and have implemented it properly in their system, and this and many other reasons are the ones that led to the realizations of this project to introduce Mexico and improve health services in the country. It is hoped that practices that are capable of giving students greater confidence when treating patients can be realized later, as well as strengthening theoretical knowledge in a simulated environment.

In the future, there will be many opportunities for virtual, augmented, and immersive reality applications to create more detailed VR interventions that can be combined with other technologies such as artificial intelligence, wearable sensors, and Big Data [26]. These technologies must be implemented in clinical practice, they may face layers of barriers, such as the cost of hardware, construction, integration of system software and hospital database, actual use of clinical equipment for used equipment and user-friendly and long-life span. The long-term use of these devices will affect the health of users, staff training in the introduction of new technologies, the skills required by the medical staff, etc., are potential problems and challenges in the future.

References

1. Gaba, D.M.: The future vision of simulation in health care. Qual. Saf. Health Care **13**(suppl1), i2–i10 (2004). https://doi.org/10.1136/qshc.2004.009878
2. Gaba, D.M.: The future vision of simulation in healthcare simulation in health-care. J. Soc. Simul. Healthcare **2**(2), 126–135 (2007). https://doi.org/10.1097/01.sih.0000258411.38212.32
3. Bell, I.H., Nicholas, J., Alvarez-Jimenez, M., Thompson, A., Valmaggia, L.: Virtual reality as a clinical tool in mental health research and practice. Dialogues Clin. Neurosci. **22**(2), 169 (2020)
4. Petrizzo, M.C., et al.: Utilization of high-fidelity simulation to address challenges with the basic science immunology education of preclinical medical students. BMC Med. Educ. **19**(1), 1–10 (2019). https://doi.org/10.1186/s12909-019-1786-5
5. Shao, X., et al.: Virtual reality technology for teaching neurosurgery of skull base tumor. BMC Med. Educ. **20**(1), 1–14 (2020). https://doi.org/10.1186/s12909-019-1911-5
6. Marone, E.M., Rinaldi, L.F.: Educational value of virtual reality for medical students: an interactive lecture on carotid stenting. J. Cardiovasc. Surg. **59**(4), 650–651 (2018)
7. Rajeswaran, P., Kesavadas, T., Jani, P., Kumar, P.: AirwayVR: virtual reality trainer for endotracheal intubation-design considerations and challenges. In: 2019 IEEE Conference on Virtual Reality and 3D User Interfaces (VR). IEEE (2019). https://doi.org/10.1109/vr.2019.8798249
8. Nursing simulation: About & resources: Healthcare simulation (2019). https://www.healthysimulation.com/nursing-simulation/
9. Villarruel, D.E.J., Guzmán, J.F.H., Dávila, M.S.M., Moreno, M.M.A., Pineda, A.d.J.T.: Aprendizaje cooperativo como estrategia didáctica en ciencias de la salud. Enfermería Investiga 1(3 Sep), 107–111 (2016)
10. Rodriguez, L.J., et al.: La simulación clínica como herramienta pedagógica. percepción de los alumnos de grado en enfermería en la ucam (universidad católica san antonio de murcia). Enfermería Global **13**(1), 175–190 (2014)
11. Deshpande, A.A., Huang, S.H.: Simulation games in engineering education: A state-of-the-art review. Comput. Appl. Eng. Educ. **19**(3), 399–410 (2009). https://doi.org/10.1002/cae.20323
12. Sattar, M.U., Palaniappan, S., Lokman, A., Hassan, A., Shah, N., Riaz, Z.: Effects of virtual reality training on medical students' learning motivation and competency. Pakistan J. Med. Sci. **35**(3), 852 (2019)
13. Akaike, M., et al.: Simulation-based medical education in clinical skills laboratory. J. Med. Invest. **59**(1, 2), 28–35 (2012). https://doi.org/10.2152/jmi.59.28
14. Contreras Olive, Y., Reyes Fournier, M., Nates Reyes, A.B., Pérez Arbolay, M.D.: Los simuladores como medios de enseñanza en la docencia médica. Revista Cubana de Medicina Militar **47**(2), 1–14 (2018)
15. DávilaCervantes, A.: Simulación en educación médica. Investig. Educ. Méd. **3**(10), 100–105 (2014)
16. Healthcare technology. https://builtin.com/healthcare-technology
17. Jones, F., Passos-Neto, C., Braghiroli, O.F.: Simulation in medical education: brief history and methodology. Principles Pract. Clin. Res. J. **1**(2), 56–63 (2015). https://doi.org/10.21801/ppcrj.2015.12.8
18. Rudolph, J.W., Raemer, D.B., Simon, R.: Establishing a safe container for learning in simulation simulation in healthcare. J. Soc. Simul. Healthcare **9**(6), 339–349 (2014). https://doi.org/10.1097/sih.0000000000000047
19. Caballero Martínez, F.: La simulación: el entorno clínico virtual. Educ. med. (Ed. impr.) pp. 12–19 (2017)

20. Mendiola, M.S., Franco, A.I.M.: Informática biomédica. Elsevier Health Sciences (2018)
21. Neira, L., Castañeda, E., Torres, C.: Use of virtual reality in the nursing school's toco-surgery teaching process. In: Ahram, T., Taiar, R., Langlois, K., Choplin, A. (eds.) IHIET 2020. AISC, vol. 1253, pp. 202–208. Springer, Cham (2021). https://doi.org/10.1007/978-3-030-55307-4_31
22. Schneiderman, B.: Designing the User Interface: Strategies for Effective Human- Computer Interaction. Pearson, Boston (2017)
23. Dam, R.F., Siang, T.Y.: What is design thinking and why is it so popular? (2018). https://www.interaction-design.org/literature/article/what-is-design-thinking-and-why-is-it-so-popular
24. Ceangal, J.K.: The relevance of video games and gaming consoles to the higher and further education learning experience (2002)
25. Wiederhold, B.K.: Lessons learned as we begin the third decade of virtual reality (2016)
26. O'Connor, S.: Virtual reality and avatars in health care (2019)

Author Index

F. Gorski et al. (Eds.): MANUFACTURING 2022, LNME, pp. 175–176, 2022.
https://doi.org/10.1007/978-3-030-99769-4

The manufacturer's authorised representative in the EU is Springer Nature Customer Service Centre GmbH, Europaplatz 3, 69115 Heidelberg, Germany. If you have any concerns regarding our products, please contact ProductSafety@springernature.com

Printed and bound by CPI Group (UK) Ltd, Croydon, CR0 4YY
30/09/2026
02225320-0001